H. H. (Harry Hubbell) Kane

Drugs that enslave

the opium, morphine, chloral and hashisch habits

H. H. (Harry Hubbell) Kane

Drugs that enslave
the opium, morphine, chloral and hashisch habits

ISBN/EAN: 9783744739474

Printed in Europe, USA, Canada, Australia, Japan

Cover: Foto ©berggeist007 / pixelio.de

More available books at **www.hansebooks.com**

LAOCOON

...

Virgil, Aeneid

...
killing the arm and body ... with the ...

Drugs That Enslave.

THE

OPIUM, MORPHINE,

CHLORAL AND HASHISCH HABITS..

BY H. H. KANE, M. D.,
NEW YORK CITY.

— — —

"They are drunken, but not with wine; they stagger, but not with strong drink."—Isaiah.

"What warre so cruelle, and what siege so sore,
To bring the sowle into captivitie,
As that fierce appetite doth fain supplie!"

PHILADELPHIA:

PRESLEY BLAKISTON,

1012 WALNUT STREET.

1881.

To

Dr. Alexander J. C. Skene,

Professor of the Medical and Surgical Diseases of Women and the
Diseases of Children, in the
LONG ISLAND COLLEGE HOSPITAL, BROOKLYN, N. Y.,

THIS LITTLE WORK IS DEDICATED,

as a mark of the high esteem,
both as regards his scientific attainments,
untiring energy, and the largeness of his heart, in which he is
held by
his old pupil and sincere admirer,

THE AUTHOR.

PREFACE.

The idea of writing this little work was first suggested to me by the numerous letters received from physicians at home and abroad, asking for information on various points connected with the symptomatology, prognosis and treatment of the various "habits." While manifesting an earnest desire to become acquainted with the main features of the conditions in question, many presented a lamentable ignorance of the simple facts relating to them.

These facts I have endeavored to present in as comprehensive, yet concise, a form as possible. But little space has been devoted to the study of the hashisch habit, owing to its rarity in this country.

The statements as to the dangers and peculiarities of these conditions, aside from what I have myself observed, are based upon the literature of the subject, and the letters of nearly a thousand correspondents in various parts of the world, to whom I take this occasion for returning my hearty thanks.

191 West Tenth Street,
New York City.

A curse that is daily spreading, that is daily rejoicing in an increased number of victims, that entangles in its hideous meshes such great men as Coleridge, De Quincey, William Blair, Robert Hall, John Randolph, and William Wilberforce, besides thousands of others whose vice is unknown, should demand of us a searching and scientific examination.

As an illustration of the enormous increase of the use of opium and morphia in the United States, the following statistics have a painful interest, and it must be remembered that this is no exceptional case. In one of our large cities, containing, twenty-five years ago, a population of 57,000, the sales of opium and morphia reached 350 pounds and 375 ounces, respectively, or about 43 grains of opium and 3 grains of morphia yearly for each individual, if the consumption was averaged. The population is now 91,000, and 3500 pounds of opium and 5500 ounces of morphia are sold annually. While the population has increased 59 per cent., the sale of opium has increased 900 per cent., and morphia 1100, or an average of 206 grains of opium and 24 grains of morphia to every inhabitant. But there are additional sales of from 400,000 to 500,000 pills of morphia, which would give us 170 ounces more of the drug. One-fourth of the opium sold is consumed in its natural state, and three-fourths are made into opiates, the principal one being laudanum.

The following is official from the New York Custom House :—

Imports of opium into the United States for ten fiscal years, ending June 30th :—

1871,	315,121 ℔s.	$1,926,915
1872,	416,864 "	2,107,341
1873,	319,134 "	1,978,502
1874,	395,909 "	2,540,228
1875,	132,541 "	939,553
1876,	388,311 "	1,805,906
1877,	349,223 "	1,788,347
1878,	430,950 "	1,874,815
1879,	405,957 "	1,809,696
1880.	533,451 "	2,786,606

Facts like these must, we think, arouse attention. They show a fearful drift. There is a worse form of intemperance than that which comes from bad liquor, although the choice would seem to be between the devil and the deep sea :

"And in the lowest depth, a lower deep,
Still threatening to devour us, opens wide."

CONTENTS.

7

CHAPTER X.

CHAPTER XI.

CHAPTER XII.

CHAPTER XIII.

THE OPIUM AND MORPHINE HABITS.

CHAPTER I.

THEIR PATHOLOGY.

A higher degree of civilization, bringing with it increased mental development among all classes, increased cares, duties and shocks, seems to have caused the habitual use of narcotics, once a comparatively rare vice among Christian nations, to have become alarmingly common.

Increase in mental strain, hot-house development of the passions, lessened physical labor and increased mental work, have been gradually giving us bodies in which the nervous element largely preponderates. Persons who may be classed under the head of "nervous temperament" are daily on the increase.

Diseases are to-day as different from diseases of a century ago as is their treatment. While the average individual now does more mental work in an hour than did our ancestors in six hours, we are not one-sixth as well able to bear the intellectual strain as they were.

Nine-tenths of us neither eat, sleep, exercise, bathe, or procreate in a proper way. It is all hurry and turmoil; little rest and much care. Generation by generation our physical natures are changing, and in the children of each succeeding generation we see the preponderance of the nervous element; a gradual evolution of that or those peculiarities most prominently brought forth by the exigencies of the individual and national life of a people.

Finding pain, "nervousness" and hysteria constantly claiming his attention, and that nothing relieves them so well as opium, or its alkaloid morphia, which is six times the parent strength, the physician resorts to their use more and more freely, expecting as soon as the more distressing symptoms pass away to pursue another and more permanent plan of treatment. The patient, however, having once experienced relief, insists upon the further use of the drug, sometimes feigns illness, in order to procure it, finally obtains some herself, and in guilty secrecy drifts rapidly into the habit.

Some physicians are weak enough to place the means of gratifying this morbid appetite directly in the hands of the patient. This is more especially the case since the hypodermic use of the drug has become common. So magical is the effect of this mode of administration, so rapid and forcible the action of the drug, that many persons will not rest content until they possess and are using the instrument. As the affections for which opium and morphine are most commonly used are chiefly found in neurasthenic patients, and as these patients are ever ready to indulge in excesses, in both stimulants and narcotics, it is not surprising that the number of victims to this slavery is daily on the increase, both in town and country. Moreover, nervous affections are on the increase: pain without any very apparent cause, nervousness from the most trivial things. Neuralgias are more common. Insanity also. Suicide is daily more frequent.

Those not acquainted with the truth in this matter will be surprised to learn that there are to-day thousands of educated and respectable people in all countries and \ among all classes, confirmed habituès; slaves to a habit that is more exacting than the hardest taskmaster, that they loathe beyond all else, and yet that binds them in chains that they are wholly unable to break.

Everything must give way to this vice. Business is neglected or but imperfectly performed; family ties are sundered; hope, ambition, happiness, self-respect are meaningless words; the one thing that fills the mind is the gratification of this passion, which they loathe, but from which they cannot break.

Thus from day to day, week to week, year to year, they go on; not living—simply existing. Each day, each hour, each minute binds them more firmly, until at last they feel their own inability to cope with the demon that has overpowered them, and abandon themselves, hopelessly, listlessly, to the vice. Repentance comes too late. The momentary pleasure, the short period of excitement, the hour of vivacity bears fruit a thousand-fold; fruit, the bitter taste of which must last them a lifetime. That which at first gave them pleasure has now become the veriest tyrant, enforcing long hours of pain and anguish, gloom and despondency. They do not continue its use *because it gives them pleasure*, but simply because it is the only thing that, in increasing doses, can save them from the torment it has itself imposed; because without it they are sunk into a. living hell. The mind is incapable of healthy action, the temper is decidedly aggravated, the person taking offence at and scolding furiously about things that in health, or while under the influence of opium, would excite no comment. / They suffer from terrible nightmares. They are constantly on the edge of imaginary precipices, or falling, falling down dizzy heights.

Sleep, if had at all, is broken, unsatisfactory and fraught with the most frightful and torturing dreams, into the warp and woof of which are constantly woven the most horrible sights. Now they are the victims of some terrible accident, again, they are hurried on by some malignant persecution. They fancy that they are drowning, that they are being burned at the stake, inhaling the sickening odor of their own burning flesh, feeling it peel from their aching bones. Then comes the awakening with a start or scream. The gradual realization that these things are not real; the cold sweat; the trembling of the limbs; the sense of utter exhaustion from which they sink into sleep once more, to live again the agonizing scenes of their diseased imagination, waking and sleeping and counting the minutes as days, the hours as years, until morning finally comes.

Nor are the torments of day much less than those of the night. The stomach rebels; nausea is persistent and distressing; saliva gathers in the mouth; there is sinking at the pit of the stomach; severe cramps of the intestines; the lips and throat are dry and parched; the tongue swollen. A dry, irritating cough sets in. Pains girdle the body and shoot with agonizing intensity down the limbs and into the face. The muscular system fails; locomotion is attended with difficulty; the sufferer staggers like a drunkard; the muscles of the face and eyelids twitch; the hands shake so that a glass falls from them, and it is impossible to pick up a small object. The circulation is affected; flushing and chilliness alternate; the eyes are dry, and feel as though filled with sand. The mind wanders; delirium supervenes; diarrhœa and vomiting set in, and sometimes collapse, and a more pitiable object can nowhere be found.

It is at this time that the sufferer, tortured beyond all

power of endurance, would sell body, soul, *anything*, to obtain that drug which, while it gives no fresh pleasure, removes these ill effects, as if by magic.

A dose is taken. A pleasant sense of warmth pervades the body; the mind clears, the hands become steady, the gait natural, the pains vanish, the nausea ~~and diarrhœa~~ ceases and existence becomes again bearable./

Each dose must be a little larger than the preceding, in order to obtain the desired effect. In some cases the increase is very slow, but these are rare exceptions. Rarer still are those instances where no increase is necessary.

I have here portrayed the suffering of one who has been using the drug for a considerable time, or for a shorter time in large doses. The chains, though not at first galling, are nevertheless there, and each succeeding dose rivets them tighter.

There are certain rare cases where opium seems, instead of doing harm, to be of positive benefit to the person using it. Dr. Joseph Parrish, a veteran observer of these cases, wrote me that he had known of several. One is related by Dr. Golding Bird.* A lady, probably hysterical, took morphia for the relief of paroxysmal pain in the loins. She had been taking it for several years. For the past two years she had increased the dose to ten grains, taken three times daily. There were no obvious ill effects; functions were properly carried on, the appetite was good, and there was no known organic disease.

The therapeutics of any epoch is strictly in conformity with the most prominent disease or symptom of the people upon whom the physicians practice. This is true of whole countries and sections of countries as well as of times.

Formerly, when it was common for physicians to pre-

* *Practitioner*, Vol. VI, p. 149.

scribe opium, it was this drug that the people ate. As morphine came into fashion, it was prescribed largely by the profession and the persons forming the habit at that time, as a rule, ate morphine. Habituès of the opium epoch also resorted to it, finding it so much more powerful than the crude drug, so much less bulky, and it did away with the necessity for calling forth a response to opium from the shattered system by resort to alcoholic stimulants. At the time in which De Quincey, Coleridge, and Southey lived, the people and the profession knew little of the opium habit, save among foreign nations; the habituès were few in number, and, as a consequence, when De Quincey's article appeared, it created a most decided impression on the public mind; an impression not yet effaced, and one which bore with it an incalculable amount of harm. Men and women who had never heard of such a thing, stimulated by curiosity, their minds filled with the vivid pictures of a state of dreamy bliss, a feeling of full content with the world and all about, tried the experiment, gradually wound themselves in the silken meshes of the fascinating net, which only too soon proved too strong to admit of breaking.

There is no question in my mind that, in writing his "Confessions," De Quincey left a large debit on the side of truth, and handed down to succeeding generations a mass of ingenious lies; more pleasantly the fiction, vaporizing from a laudanum-soaked brain. He must needs seek some justification for his life of willful misery, for the blasted hopes, ambitions and prospects of what might have been a noble career, and he offered the dream life, the fuller development of benevolence, and the many pleasures so fantastically portrayed, as a justification, in part at least, for his sin.

Nor does the final confession of the intense pain, the

abject misery, the tottering of the mind, the crumbling of the reasoning and will power, and the ever attendant and impenetrable gloom of a living hell, serve to fully counteract the baneful effects of the portrayal of the pleasures of opium. The reader, confident of his ability to stop short of the ever-shifting line that divides the happiness from the misery, is in no wise deterred from trying the danger-fraught experiment. I know of several patients who began the use of opium simply from reading this most pernicious book.

Upon persons living in temperate and cold climates this drug does not have any such effect, with reference to the subject matter of dreams, as upon Orientals. Indolent, over-fed, and by reason of their mode of life, religious associations and habits of thought, fancifully imaginative, it is not surprising that they should enjoy, while under the influence of the drug, grotesque, and to them, pleasant, dreams. Did the opium cause dreams foreign to the picture daily conjured by their fertile imaginations, it would indeed be more surprising. That it does not produce such effects on our plain, work-a-day people is not to be wondered at.

As I have already said, the preparation of the drug used and the manner of using it in any epoch has been exactly in consonance with the practice of physicians at that time. Of late years physicians are becoming more and more addicted to the subcutaneous use of morphine, and as a consequence, the number of persons who habitually use the drug in this way is daily on the increase. Eight-tenths of those from whom I hear and of those who come to me for treatment are using the drug subcutaneously.

Dr. Charles Warrington Earle, of Chicago, in a very able

and well written little pamphlet,* is of the opinion that the
majority of habituès do not use the drug in this way. In reply
I can only reassert my opinion just expressed, and must say
that the tendency of these patients to falsify, and their
delicacy in disclosing their manner of using the drug to
the druggists from whom they obtain their supply, must
be taken into consideration. Dr. Earle bases his conclusions on 235 cases, the histories of which have chiefly been
obtained from druggists.

Be it understood, however, that I do not maintain that
the majority of opium and morphine takers use the latter
drug by the hypodermic syringe. I simply say that in
my experience this manner of using the drug is largely on
the increase among habituès, and will go on increasing
from year to year, in the same manner that morphine is
rapidly replacing opium in the practice of physicians.
This is well shown by one of Dr. Earle's carefully prepared
tables : —

KIND OF NARCOTIC.

Morphia was used in	. . . .	120 cases.
Tincture of opium was used in	. .	30 "
Paregoric " "		5 "
McMunn's elixir " " .		2 "
Gum opium " "	.	50 "
Dover's powders " "	. .	1 "
Unknown " "	. .	27 "

235

The age at which this habit is most common is from
thirty to forty, both in males and females. The following
table, which, as Dr. Earle states, is only approximative, is
of interest in this connection : —

* " The Opium Habit; " reprint from the Chicago *Medical Review*, October and
November 5th, 1880.

Males—

From 20 to 30 years	5
From 30 to 40 years	19
From 40 to 50 years	11
From 50 to 60 years	7
From 60 to 70 years	1
From 70 to 80 years	1
Unknown age	22
Total	66

Females—

From 10 to 20 years	2
From 20 to 30 years	18
From 30 to 40 years	39
From 40 to 50 years	22
From 50 to 60 years	14
From 60 to 70 years	4
One-third entire number prostitutes, probably from 15 to 50	56
Unknown age	14
Total	169

Females are more frequent victims than males, in the proportion of three to one. This is undoubtedly due to the fact that women more often than men are afflicted with diseases of a nervous character, in which narcotic remedies are used sometimes for a long period, and also to the fact that in some instances it is used by them in place of alcoholic stimulants, its effects being less noticeable and degrading, although none the less intoxicating.

Both males and females are usually of the higher orders, in point of intellect and culture. In some cases business failure or family trouble has been the incentive for a resort to the use of the drug. In some instances the fact that opium eating had ruined the mental powers of the victim,

or caused him to be careless or negligent of his home relations, has led to the business failure, or the sundering of
family ties. The majority of patients come from the
middle classes, those people who are continually toiling
and worrying in the almost ceaseless endeavor to "keep
up appearances."

The fact that most opium eaters are married, widow or
widower, is probably explainable on the ground that in
the majority of instances, the patients among whom it is
most common are at just the age when marriage has taken
place. In some the habit is contracted before, in others
after marriage.

I knew of one example where the wife, a young woman
of eighteen, contracted the habit of using the drug subcutaneously, through the carelessness of her physician. The
husband began then to use it himself, and to-day the two
are separated, the wife partially insane, the husband a confirmed habituè and also an alcoholic drunkard. One who
sees much of this disease meets with some very sad cases.

CHAPTER II.

PREPARATIONS EMPLOYED. MANNER OF USING.

The following are the various preparations used by opium and morphia takers—

Crude Opium.—Variable in strength.

Tincture of Opium, Laudanum.—1 grain of opium to every 25 drops, or 1 grain of opium to every 13 measured drops, *minims.*

Tincture of opium, Camphorated (Paregoric).—1 grain to the ounce, or 480 drops.

McMunn's Elixir.—Same strength as laudanum.

Dover's Powder.—1 grain of opium and 1 grain of ipecac to every 10 grains of the powder.

Sulphate of Morphia. ⎫ Salts of opium, 1 grain of either
Acetate of Morphia. ⎬ being equal to about 6 grains
⎭ of opium.

Laudanum was De Quincey's favorite. He avers that he has taken as high as 8000 drops in the course of twenty-four hours. Coleridge is credited with having taken more. A patient, a lady, whom I saw in consultation with Dr. Brennan, of this city, took a half pint every morning. Without it she was totally unfit to converse or to be seen. Many of the old fashioned habitués use the gum opium, but few, paregoric, and a less number still, Dover's powder. In order to get a sufficient amount of opium for the average user, from Dover's powder, the ipecac, which is present in equal amount, would produce nausea, and offer a further bar to its employment. Pulverized opium and extract of opium have been used in the form of supposito-ries. This was the case with one of my patients, a gentleman suffering from chronic inflammation of the bladder ; after a time he abandoned the opium, using in its stead morphia, in gradually increasing doses.

I have already given Dr. Earle's table as to the relative frequency with which the different preparations are used. I here append his table of quantities. Of this and the preparation sold, druggists from whom he obtained his statistics would be able to judge with a reasonable degree of accuracy.

QUANTITY.

Morphia—

21 persons use from 1 to 3 grains each day.
17 " " 3 to 6 " "
12 " " 6 to 10 " "
10 " " 10 to 15 " "
12 " " 15 to 20 " "
7 " ½ a drachm " "
6 " 1 drachm " "
20 " 1 bottle per week.
5 " 2 " "
1 " 11 bottles each month.

Tr. Opium—

15 persons use 1 drachm each day.
4 " 3 " "
7 " 4 " "
12 " 1 ounce "
4 " 2 " "
1 " 3 " "
1 " 4 " "

Gum Opium—

3 persons use 10 grains each day.
5 " 20 " "
9 " ½ drachm "
12 " 1 " "
4 " 2 " "
2 " 3 " "
1 " 4 " "

Morphia is used either by the mouth, rectum or hypodermically. I know of a curious case, to be detailed more fully at a subsequent time, where the patient, a young lady, took morphine first subcutaneously, then by the rectum, and finally by the mouth.

I am of the opinion that morphia, when continuously used, works more harm when taken by the mouth than when taken subcutaneously. Moreover, that the habit is more difficult to break, and relapses more prone to occur, in the former than in the latter class of cases. Opium or morphine by the mouth, after a time, exercises a directly paralyzing and anæsthetic effect upon the mucous membrane of the alimentary canal, and gives rise to hepatic derangement and sometimes serious disease. The subcutaneous injection of morphia undoubtedly exercises a baneful effect upon the same surfaces and organs, but not so directly as when taken by the mouth.

Among the Eastern nations opium is used chiefly by smoking. As this is rarely seen in this country, save among people from those countries, it does not especially concern us. Moreover, a full account of these practices may be found in books of travel and other treatises upon that special branch of the subject.

To Dr. Alexander Wood, of Edinburgh, belongs the credit of first discovering and practically applying the hypodermic syringe to the treatment of painful affections, more especially neuralgias. Drs. Isaac E. Taylor and Washington, of this city, claim to have used it in this country in 1839, while Wood's discovery was made in 1843, and his first article appeared in 1855. Wood's instrument was first brought to this country in 1856, by Prof. Fordyce Barker, of this city. Then comparatively unknown, it is now in the possession of almost every physician in the

world. In my work upon the hypodermic use of morphia*
I made, in the preface, the following statement, the belief
in which has since been strengthened : " There is no proceeding in medicine that has become so rapidly popular ;
no method of allaying pain so prompt in its action and
permanent in its effect ; no plan of medication that has
been so carelessly used and thoroughly abused ; and no
therapeutic discovery that has been so great a blessing and
so great a curse to mankind, as the hypodermic injection
of morphia."

The danger of forming the habit from the use of the
drug in this way is undoubted. Correspondents, physicians from all parts of this country, England and France,
assert this and detail cases. Levenstein† gives many instances in Germany, and Dr. Loose, of Bremen, sends me
the report of an able paper read by him before a medical
society, wherein he cites cases, deplores the rapid spread
of the practice, and advises special legislation upon the
subject. He remarks for that country what Dr. J. B.
Mattison, myself and others have noted in the United
States, i. e., that many of the victims are members of the
medical profession, in good standing. One hundred and
thirty-one physicians report to me one hundred and eighty-
four cases of the morphia habit, in all of which it was
contracted by the use of the hypodermic syringe.

The largest amount taken in the twenty-four hours by
this method is reported to me by Dr. W. F. McAllister,
of the Quarantine Office, San Francisco, California : "A
physician, resident in this city, contracted dysentery in
Shanghai, China, in the summer of 1873. Morphia was
used hypodermically, and he drifted into the habit of using

---

* " The Hypodermic Injection of Morphia. Its History, Advantages and Dangers."
N. Y., 1880.

† " Morbid Craving for Morphia," London, 1878.

Fig. 1.

FIG. 1.—Hypodermic Syringe. Glass cylinder with metal casing. Indexed on glass.

Fig. 2.

FIG. 2.—Plain Glass Syringe, with needle. Index on stem.

Fig. 3.

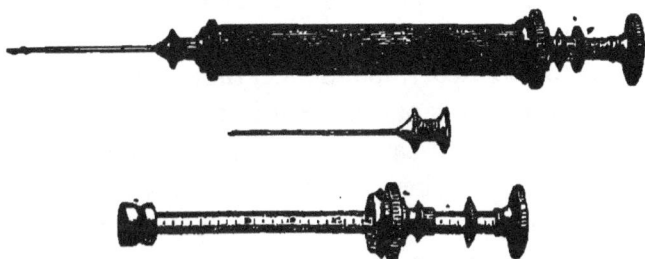

FIG. 3.—Plain Metal Syringe, graduated on stem.

Fig. 4.

FIG. 4.—Plain Metallic Syringe, graduated on stem, with a cap for securely closing needle end of barrel, and a device by which needle can be carried in hollow stem. This is an excellent instrument. All those here shown were made by Codman & Shurtleff, of Boston, and are durable, accurate and easily cleaned.

the drug in this way himself; the habit resulting in his death in 1878. I was called to see him professionally in 1875. He was at that time residing in Hong Kong, China. He was consuming each day *seventy-two grains* of the sulphate of morphia in three doses: twenty-four grains to the dose. This he continued to do until the day of his death."

Prof. T. Gaillard Thomas, of this city, relates to me two fatal cases, in young persons, death being due to extreme exhaustion, dependent on imperfect nutrition and a gradual depression of the nervous and circulatory systems.

In consequence of the skepticism evinced by some physicians as to the danger of contracting the habit in this way, I feel myself called upon to urge upon them the utmost care in the use of this instrument, especially in the case of neurasthenic patients. I have had so many deplorable cases of this kind brought to my notice, either as patients or through correspondence, that I cannot help but feel that this matter is one of great importance, and worthy of more attention and care than is usually given it.

I have already spoken of the two ways in which the opium or morphia habit is formed. A patient of a nervous temperament, suffering from some painful or spasmodic disease, is attended by a physician, who administers one of these drugs, by the mouth, rectum, or subcutaneously. The relief to suffering is magical, and persists so long as the effect of the drug lasts. With a return of the pain comes the natural desire to have it relieved, and as the narcotics just spoken of have been found especially efficacious the patient desires its repetition. This may go on for weeks and months, until the disease, of which the pain was but a symptom, is cured, or it may never be cured. At any rate, the patient awakes to the knowledge that he must have his narcotic. Life without it is unbearable, and instead of putting forth an extreme effort of the will, as is done by some, and then and there abandoning its use, the majority allow themselves to drift into this habit of daily intoxication with opium. The majority of habituès first use the drug for the relief of pain, and then find themselves unable to break loose from it. Some seem to be so constituted that a single or a few doses of drugs of this nature light up in them an irresistible desire to continue their use. There seems to be a morbid craving for *something*, exactly what is not known, until the narcotic is tried, when this morbid appetite is satisfied and fixed, and becomes the typical "morbid craving for morphia." It will be dipsomania, morphia-mania, chloral-mania, hasheesh-mania, according as the one or other drug is presented to the patient in the condition of craving. Some persons are undoubtedly born with, and some acquire, this craving for some narcotic or stimulant.

A person knowing this fact can readily see how like putting a match to gunpowder it is to give these people opium or morphine in any form, and how inevitably the reading such a book as that of De Quincey's would create a longing and open the way to a road that has a certain ending in a life's bondage. Such as these are to be pitied, and deserve the kindest treatment and the most judicious care. They are like a person who has lost a limb, or is suffering from a cancer. In the one instance they lack a certain something that should be there, and which is necessary to the free, full and proper enjoyment of life; in the other case—the acquired tendency or craving—they have a pernicious addition to the system that threatens them with death.

Pity, then, rather than blame, at the same time using every legitimate means to break up both the habit and the tendency.

3

CHAPTER III.

GENERAL SYMPTOMS CLASSIFIED AND ANALYZED.

I am of the opinion that a clearer and consequently more thorough understanding of this part of the subject can be obtained by classing the various effects of the continued use of morphine or opium under the headings of the different systems and apparatuses, and have, therefore, adopted this method of presenting these matters to the reader.

THE DIGESTIVE APPARATUS.

(Stomach, Liver and Bowels.)

The first effect of opium or its alkaloid morphia on the stomach is, in the majority of instances, to increase the appetite and cause a pleasant sensation of warmth and a feeling of general contentment. After a time, however, especially if the drug is taken by the mouth, the appetite is materially impaired, the bowels become constipated, there is a sensation of sinking at the pit of the stomach, the circulation through the liver is interfered with, a catarrhal inflammation of the small intestine and stomach supervenes, the opening of the bile duct is partially obstructed by the swollen mucous membrane, and the result is a more or less decided staining of the skin with the biliary coloring matter. This is due, in part, to interference with the cells in the proper performance of their functions, undoubtedly through the medium of the nerves passing to them. At the same time, the gastric juice is diminished in amount and lessened in strength, as is also the case regarding the intestinal and pancreatic secretions, the digestion of food is imperfectly performed, and in .consequence, a condition of malnutrition results, showing itself in wasting of the muscles and loss of the fatty tissue beneath the skin, especially of the face, breast and abdomen. As the gastric catarrh increases, nausea and finally vomiting ensue, still further debilitating the patient. Derangements of the stomach, liver and bowels are much more common among those who use the drug by the mouth. I have, however, seen patients who, while using morphia hypodermically, have suffered intensely with jaundice, catarrh of the stomach, nausea and vomiting. When these disorders do arise in the course of the subcutaneous use of morphia they are prone to be very acute in character. A gentleman, aged about 32, came under my notice some months ago, suffering from dropsy of both legs, which, from the middle of the leg to and about the ankles, were covered with collections of minute red spots, evidently points at which the capillary blood vessels had ruptured. An examination of his urine discovered thirty per cent. of albumen and some tube casts from the kidneys. Also a decided amount of morphine, and considerable blood. He was put upon tonic remedies ; the legs, previously dressed with sulphate of iron ointment, were tightly bandaged. After about two weeks the dropsy lessened, the spots disappeared and the general health improved sufficiently to allow of a business trip to the West. No sooner had he neared his destination, however, than the dropsy of the legs again appeared, and the old trouble came back in full force. In addition, he became somewhat jaundiced, water made its appearance in the abdomen, and a most distressing nausea harassed him day and night. During all this time he managed to do an unusual amount of literary work, that required both knowledge and tact. On his return to New York, he was in a very weak and debilitated condition. The stomach rebelled at almost every kind of food, the belly was distended with water, the nausea intense, the vomiting frequent, and urine high colored, bloody and containing about sixty per cent. of albumen. While in the West an attempt was made to gradually diminish the amount of morphia, but owing to severe illness, was abandoned. Nor did I think it advisable to try again until the general health was better. Upon iron, in different forms (dialyzed, Bland's pills, and muriated tincture), strychnia, phosphorus and gentian, he gradually improved, and is now just commencing the treatment by gradual reduction. He consumes daily about ten grains, using it subcutaneously. He contracted the habit some five years ago, through its being administered to him hypodermically, during a severe attack of acute articular rheumatism. Two years ago he was broken of the habit by a physician in this city, who pursued the plan of sudden deprivation, with the result of nearly killing the patient.

While traveling in Germany, for the purpose of reporting certain facts for his paper, he was suddenly taken with congestion of the brain, accompanied by intense pain and a state of semi-consciousness. A German physician who was called in, not knowing the man's previous history, gave him a subcutaneous injection of morphia, with the result of re-establishing the habit.

Fluid in the abdomen is not a common complication, and when it does occur, crowds the stomach upward, interferes with the circulation of blood in the viscera in the abdominal cavity and renders the nausea and vomiting still more troublesome.

The stools voided by morphia-maniacs are sometimes clay-colored, being devoid of biliary coloring-matter. The absence of bile in the intestinal canal allows of decomposition of the food, with the production of ill-smelling gases which regurgitate through the stomach, or pass by the bowels. Constipation is the rule. Sometimes it alternates with severe diarrhœa. In one patient whom I saw, the fæces collected day by day in the bowels, until impaction resulted and the mass was removed, with the greatest difficulty, by means of large enemata and the handle of a spoon. This lady had been using the drug, crude opium, for only eight months.

This constipation from hardened condition of the fæces produces irritation, intestinal catarrh, and often hemorrhoids accompanied by an eczematous eruption about the anus.

Heartburn, not from an excess of gastric juice, as is so often supposed, but from an acid fermentation of the food, is not an uncommon symptom.

Some patients go on for years with no disturbance of the bowels, liver and stomach, beyond a slight catarrh and some jaundice, but these troubles are sure to come sooner or later.

In many instances as the appetite for food decreases the appetite for alcoholic stimulants replaces it. A young lady, of twenty, formerly a patient of mine, after using the drug subcutaneously for three years, lost her appetite and subsisted for nearly eight months upon lager beer, a few crackers and a little toast. She would consume about one gallon of beer a day. During this time she somewhat diminished the amount of morphine used (eight grains, dropped to six grains, in twenty-four hours) and gained in weight and health. At the end of this time, however, severe gastric disturbance supervened, and she lay for months in a precarious condition. She is now cured of the habit.

Alcoholic stimulants are used by some habitués to increase the effect of the morphia, which it does to a certain extent.   They delude themselves into the belief that they will be enabled to thus reduce the amount of morphia used, but in the majority of cases this is a fallacy.   Those who prefer laudanum and paregoric to opium or morphine often do so on account of the amount of alcohol contained in these preparations.   I knew a gentleman, some years ago, now dead, an Episcopalian minister, who became addicted to the use of alcoholic stimulants.   His excesses were so frequent and so degrading that it was found necessary to expel him from the church.   Sobered by the blow, yet unwilling to entirely abandon his habit, he seized upon the expedient of using large quantities of laudanum, thus getting the effect of both opium and alcohol.   He died some two years after, from taking an overdose, at the end of a spree, the first and last one in the two years.   His tombstone is a monument to blasted hopes, unsatisfied ambition, a life ill-spent, and points a very pregnant moral to those who knew him.   Let a man count well the cost before he takes service with a master who is only satisfied with a blasted life, here and hereafter.

THE CUTANEOUS SURFACE.

The effects of these drugs upon the skin may be divided into those relating to *color*, *sensation* and *nutrition*. The sallow, cadaveric hue so often seen in advanced cases is partly due to discoloration by the biliary matters, which are not properly excreted by the liver, and partly to a condition of malnutrition, so often seen in persons suffering from cancerous disease.   In case there is serious interference with the working of the kidneys, the sallow color becomes less distinct, diluted expresses it well, and it has an appearance more like that of old white marble. In this case the tissue feels doughy and "soggy," and at points a pit made in the skin by the pressure of the finger remains. There is also puffiness under the eyelids and the eyes are bleared and watery. When there is no dropsy the skin is usually dry and harsh, although at times bathed with a cold, exhausting perspiration, the odor of which is oftentimes very offensive. A morphia or opium taker, previously ill-nourished, sometimes seems to improve, taking on flesh rapidly. The appearance, often best seen in the face, is very deceptive, and is, in the majority of instances, due to slight dropsy. It is sometimes seen a few weeks after these patients commence to use alcoholic stimulants, especially beer. It soon passes away, leaving them looking thinner and more miserable than before.

In those cases where the habit has not progressed very far, the skin is usually made redder and coarser than . natural, and is dotted here and there with pustules.

In one case that I have recently had under my care, that of a young lady who had never used stimulants and who had only used morphia, and that subcutaneously, for seventeen months, a swollen and reddened condition of the nose, exactly like that seen in old drunkards, was present. Two weeks after the morphine was stopped the nose began to lessen in size and the color to disappear, and the condition was entirely cured at the end of three weeks. No local applications were used.

Loss of elasticity of the skin is a common change. That nutrition is seriously interfered with is seen by this alteration, the loss of adipose tissue, the ill-balanced circulation, giving at one time coldness and paleness, at another flushing and undue warmth, and the variable secretion. A plump and well nourished person who commences the use of morphia loses those outlines of health caused by the proper development of the muscular system and normal distribution of fat beneath the skin. Face and form are both changed, and certainly for the worse.

These changes are undoubtedly due to a double action of morphine ; that on the alimentary canal preventing the proper digestion and assimilation of food, and that upon the nerves regulating the supply of blood that passes to the skin.

Eruptions after a time appear upon various parts of the body, more especially the face, chest and back. They are usually pustular, with hard, indurated bases ; sometimes papular. Herpes zoster or "shingles" is occasionally produced, and is attended by most intolerable itching ; less often purpura hemorrhagica. In the case of the literary gentleman already spoken of this was present, as was also bleeding from the kidneys. When it does occur the system must be in a greatly exhausted and the blood in a very poor condition, a condition which it usually takes a long time or large doses of the drug to produce.

Another symptom dependent on disordered nutrition of the skin is the rapid or slow falling of the hair that occurs in some patients. However much hair restorers and scalp tonics may be used in these cases, the hair, once having commenced to fall, will continue to do so until the general health improves. When the drug has been stopped a free falling out of the hair occurs, followed by a new and luxuriant growth. Indeed, sweeping out and rebuilding seems to be the rule in every tissue. Change in color of the hair has been noted in some cases.

Those affections of the skin due to the use of dirty solutions and unclean needles when the drug is given subcutaneously will be spoken of fully presently.

Sensation is variously altered. Sometimes there is a condition of hyperæsthesia, the least touch being intensely painful, the contact of the clothing causing decided irritation. Occasionally a whole limb or part of a limb or portions of the trunk or face will feel "numb or dead." Sometimes there are a series of pricking or tickling sensations that are very aggravating, as they and the numbness are often looked upon by patients as precursors of paralysis. Sensitiveness to cold is often extreme. This and the very hard clay-colored stools are also seen in saccharine diabetes, and in every case where I have noted them sugar was found in greater or less quantity in the urine of the patient. What the connection is I cannot say, but it is worthy of further study.

THE SEXUAL ORGANS.

The continued use of opium or morphine has a decided effect upon the sexual apparatus. One of the first changes to be noticed in women is scantiness, then irregularity, and finally, total cessation of menstruation. In some cases an occasional period is observed, but as a rule, the stopping is decided and permanent until the habit is abandoned. Occasionally menstruation occurs when a change is made from one preparation to another, or from one mode of administering to another.

In one of my patients, not under treatment for that, but still an habitué, menstruation showed itself, after a three years' cessation, on changing from the use of laudanum by the mouth to morphine by the skin. It occurred, however, but the once.

In the case of the young married lady already referred to, when a change was made from subcutaneous to rectal injections the flow was re-established and showed itself, although irregularly, for nearly a year.

Levenstein, who has made some interesting and valuable researches and experiments upon this matter, believes that this abnormal menopause is due to the inactivity of the ovaries. He says :—

" According to Pflueger's theory, in cases of amenorrhœa due to morbid craving for morphia, the growth of the ovarian cells would be stopped from one monthly period to another, and consequently there would be a want of stimulus on the ovarian nerves, causing on the one hand the rupture of Graaf's follicles, and producing, on the other hand, a congested state of the generative organs by reflex action. Hence, the morphia would act in the same manner on the ovaries as on other secreting glands, *i. e.*, would render them devoid of function under its continued influence. It is likely, therefore, that the menstrual discharge does not show on account of no ovulation taking place ; this also would account for the sterility."*

Accompanying this condition of non-menstruation there is always sterility. That this is due to the habitual use of morphia is proven by the fact that these women have borne a child or children before the use of the drug was commenced, and have again become pregnant and gone to full term after the habit was abandoned. So long, however, as menstruation does occur, there is a possibility of the woman's becoming pregnant.

In the majority of instances sensation is finally lost, although it is usually increased during the first few months' use of the drug. A patient of mine, a lady who had used morphine by the mouth for sixteen years, found her virile power during and at the end of that time in no way impaired. If anything, it was increased. Her dose of morphine was ten grains. She had not menstruated for a

* Levenstein, op. cit., p. 69.

long time. This was the only case in which I tried Levenstein's plan of at once stopping the drug, and I shall
certainly never try it again.

With the suppression of the menstruation, there sometimes come the usual symptoms attending the suppression
of this function in non-habituès.

Levenstein states that ladies suffering from leucorrhœa
are often cured of this complaint by the habitual use of
morphia ; the discharge returning, however, as soon as the
habit is broken, and causing labor-like pains.

Certain it is that most female patients, on breaking the
habit, suffer from a severe leucorrhœa, whether or not they
had such a discharge previous to commencing the use of
the drug. One of my lady patients who used six grains a
day, subcutaneously, had a leucorrhœal discharge before
she began the use of morphia, and upon which a four years'
habituation had no curative or modifying effect. As soon,
however, as the habit was broken, the discharge increased
greatly in quantity, and changed in character, becoming
more tenacious.

Spasmodic closure of the mouth of the vagina on
attempting to introduce the finger or a speculum, I have
noticed in two cases. It, also, readily passes away after the
use of morphia is abandoned.

A woman becoming pregnant in the early course of her
addiction to this habit will, in the majority of cases, abort
before reaching full term, especially if the amount used is
large. Levenstein has noted the fact that wives of men
addicted to the habitual use of this drug in large doses
had in the last two years never carried children to full
term, although young, healthy, and having borne children
before the husband became an habituè.

There is no question in my mind but that the excessive

use of this drug by one or both parents, but especially the mother, in case she is able to carry her child to full term, will modify disadvantageously the physical, mental, or moral development of the child thus born. A physician from the South tells me of the case of a lady who commenced the use of opium at the beginning of her pregnancy. She was delivered of a fair-sized child that grew up in fair, though not robust health, and menstruated at the proper age. She is, however, very simple and childish, still plays with dolls, although a young lady, is very eccentric, and shuns the society of young men. The mother, who is still living, ceased to menstruate at the age of 30 and has never menstruated since, some twenty years. For the past two or three years she has been using sixty grains of gum opium daily.

Dr. Alonzo Calkins[*] relates several cases where the children of such parents were either physically or intellectually deficient. I give but one of them: " At an inquest held by Dr. Macnish it appeared that a child five years of age, though to appearance only so many weeks old, had never been able to walk, nor so much as utter an articulate sound. The mother, during her gestation (as was in evidence), had taken to morphine, using a drachm a day in the months just preceding her demise. A child born before the habit had become fixed showed a normal development and the aspect of general health." Another case is related where both parents were " healthy and robust by original constitution and by habits of life too, with the exception that the woman had, for a very considerable period, been in the practice of using morphine regularly and to great excess. An infant born subject to the liabilities, had only a very imperfect physical organization, with weak intellectual indications."

By carefully conducted experiments on animals (preg-

* "Opium and the Opium Habit." Philadelphia, 1871. p. 117.

nant dogs and rabbits), Levenstein found that the continued use of morphia invariably produced abortion, the fœtus being born dead.

With the cessation of the menses the breasts usually dwindle in size, and the voice attains a more masculine tone.

In man the first indication of an effect on the sexual organs is increased desire ; this, however, giving way sooner or later to partial or total impotence. The fact, noted by Levenstein, regarding the rarity of conception and parturition in the wives of habituès, has already been spoken of. If the amount of opium or morphine used is small, impotence may not come for a number of years. The power of partial or full erection, without emission, is preserved by some. In those cases where there is inability to impregnate the female, there undoubtedly exists a deterioration of the functional power of the testes. In one case which I saw the testicles were markedly atrophied. In two cases I was fortunate to be able to examine the seminal fluid, a few days after the habit had been broken. These patients, as is usual on recovery, were troubled with erections and nocturnal emissions. In both instances the zoosperms were small, and present in less than the normal quantity. In one case, where it was possible, owing to the patient's having been previously instructed, I was able to examine this fluid once or twice weekly for nearly two months. Each succeeding examination showed a larger number of zoosperms, each sample apparently better developed than the preceding.

Even large doses sometimes fail to produce impotence. One young man, twenty-five years of age, who has been taking, subcutaneously, ten grains of morphia, for over three years, is a confirmed masturbator, and seems to have frequent emissions.

CHAPTER IV.

GENERAL SYMPTOMS CLASSIFIED AND ANALYZED.

THE URINARY ORGANS.

Albuminuria, usually temporary, is not an uncommon result of the prolonged use of morphia. The deposit, after applying heat and nitric acid, may vary from one-twentieth to one-seventieth per cent. of bulk. In some cases it comes and goes, one day appearing as a slight, hazy cloud, and on another as a measurable deposit. Albuminuria and diabetes are more commonly found in patients who use the drug hypodermically. In some rare instances casts, epithelial, granular, hyaline and bloody, are to be found in the urine, as also free renal cells, apparently healthy. This was the case with the literary gentlemen spoken of in the second chapter. This does not indicate organic disease of the kidney, for after the withdrawal of the drug the albumen and casts gradually disappear.

The specific gravity of the urine varies according to the bodily condition of the patient, the weather and the amount of fluid ingesta used. If sugar is present the gravity is high ; if albumen, *usually* low. In nervous and hysterical women there is a low gravity with an excess of the alkaline phosphates. In nearly all cases the uric acid is increased after the drug has been used for a time, the urea is very materially diminished, due probably to the small amount of food taken, the congested and deranged condition of the liver, and the impeded tissue metamorphosis. The chlorides are always slightly, sometimes markedly, diminished in amount.

Levenstein concludes, from his observations on patients and experiments on animals, in whom the continued use of morphia produced albuminuria, that this affection is due to varying blood pressure in the renal vessels through the nerves supplying them.

Contrary to the experience of Levenstein, I have found sugar in the urine of some (four) of my patients. This, like the albuminuria, was permanent in one case, and came and went in the other three. Its presence was undoubted, it reducing the copper of Fehling's liquor (previously tested) and answering to the fermentation and bismuth tests. In three of the cases it disappeared after stopping the morphia. The fourth patient is still under treatment.

Levenstein finds that *acute* poisoning by morphia, in men and animals, is always accompanied by sugar in the urine.

Strangury and retention of urine is occasionally the result of the long continued use of the drug.

The albuminuria of morphia-mania is sometimes accompanied by dropsy of the feet and limbs, less often of the abdomen and pleural sacs. When thus occurring the case is a grave one, and it will take time to decide whether there is not real organic disease of the kidneys, in which case the breaking of the habit may cause death by convulsions.

THE EYES.

It is a rare thing to find actual disease of the eyes that can be traced directly to the abuse of morphia. It is usual to find the retina somewhat congested. Blurring of sight and double vision are sometimes complained of. Muscæ volantes, or specks floating before the eyes, are sometimes seen. One patient of mine who had been near-sighted four years, during which time she had used the drug subcutaneously, claims to have regained full power of sight since the habit has been broken.

The pupils are, as a rule, contracted and regular, occasionally of normal size and irregular.

THE MUSCULAR SVSTEM.

Locomotion is rarely affected during the continuance of the habit, save from weakness. Twitching of isolated muscles, such as the orbicularis palpebrarum of one eye, is sometimes seen. When the doses of opium or morphine become very large, co-ordination is sometimes interfered with. Wasting of the muscles is found in advanced cases, as also is trembling of the hands.

RESPIRATION.

Respiration is rarely affected. In some cases there is shortness of breath on walking fast or going up a long flight of stairs. A low grade bronchitis sometimes exists, as also a short, hacking cough, that seems to come chiefly from irritation in the throat.

THE CIRCULATION.

Circulation is often affected but chiefly through the agency of the nervous system ; witness the flushing of the face, and flashes of heat over the body, followed by a cold, exhausting sweat. There is often irregular and weakened action of the heart, and congestion of the brain. The albuminuria already spoken of is due, in all likelihood, to an affection of the vessels of the kidneys, through the nerves supplying them. The blood itself, owing to a general condition of malnutrition and imperfect digestion, is unquestionably deteriorated. In some cases the vessels rupture, giving us purpuric spots on the body and hemorrhage from the kidneys and bowels. Dropsy, which sometimes occurs, is due, in the majority of instances, to variable blood pressure, and a diseased condition of the walls of the vessels, permitting the easy transudition of the watery element of the blood. Headache, flushing of the face, flashes of heat and the like, from suppression of the menses, is seen during the early period of morphia addiction, but later nothing but the nervous symptoms attendant upon this condition are manifested.

THE MIND.

Early in the use of morphia the effects upon the mind are simply those of pleasant exhilaration, a feeling of perfect contentment, good will toward all, increased conversational power and stimulation of the imaginative faculties. Sleep is preceded by a period of luxurious drowsiness, fertile in pleasant retrospective and magnificent anticipation. There is total banishment of pain and care. It is, indeed, the ideal life of a dreamer, moulded and modified according to the temperament and intellectual tendencies of the individual.

As time advances, however, the duration of these periods is found to become shorter, and it is necessary, in order to obtain the same pleasant result, to increase the amount of the drug. It has, also, to be taken oftener. Sleep, when it comes, is less profound, the dreams not so pleasant. Should the patient pass the accustomed time for the drug, the loss is at once felt and the first symptoms of rebellion show themselves. A further and larger dose is taken, to quiet the rebellious demon that rules them, and again comes the pleasure, though not so satisfying as at first. The hours of freedom from the tyrant become shorter and shorter, sleep refuses to come, grave doubts fill the mind, the temper is no longer even and pleasant, but irritable and capricious, pains show themselves in various parts of the body, the nights are long hours of torment, conversation becomes a burden, suspicion shows itself, a desire to be alone is overpowering, better feelings are blunted, benevolence is replaced by selfishness, mental stimulation and exhilarance by torpor.   Friends are neglected, the books that were once interesting are no longer so, amusements pall upon the taste, family ties, once so pleasant, are become burdensome, and life a dreary space, marked only by the hour at which the drug is to be taken.   The days are filled with repentance, the hours garnished with resolves no sooner made than broken,

"O woeful impotence of weak resolve,"

the nights years of misery and anguish, teeming with horrors beyond the power of tongue or pen to paint. Here is the plaint of one, now freed from his bondage : *   "The morphia victim dwells, after the first exhilaration is gone, in a realm of phantoms and shadows.   I saw sights more terrible than can be imagined. I felt pains that do not belong to any mortal lesion.   I have shrieked my terror, but the shriek only awoke a myriad of devils, who had been sleeping till then unseen by me.   Four months of morphia addiction sufficed to bring me to this land of horrors, where no joy came or has come since the making of the world.   My days were spent in self-indulgences.   Alone in my office, in an easy chair, I could, with poetry and interesting therapeutical works, manage quite comfortably to pass the hours away.   But let a patient summon me away from home, and my gloom and despondency was almost insupportable.   I was tormented by continual self-conflict.   Conscious of the weakness of my efforts to emancipate myself, I kept on planning some

* The Personal Experiences of an Ex-opium Habitué.—N. Y. *Medical Record*, p. 399, Vol. xiii.

new mode of attack, in the nerveless hope that I could defeat the Lethean devil whose thews were strong as steel, and yet I knew, as day followed day, and week followed week, in so far as all this mental warfare was concerned, it could bring me no help in my awful bondage.

" No dark imagery can paint the encompassing horrors of those nights of torment that belonged to the last two months of my twelve months' morphia addiction.   Not one hour that I passed in bed between midnight and noon did I know normal sleep.   In dreams that seemed more vivid than reality, I entered gloomy caves, and walked for hours over rotten cadavers, sometimes forced to step on them and be overwhelmed with loathsome odors.   I saw faces in the weird darkness, sometimes a thousand at once, and each was made of blood-red flame ; they flashed and went out.   My nightmared brain was chased and haunted by everything that can exist in a vast hell of phantoms."

Apropos, the following from Moore's " Veiled Prophet of Khorassan ":—

" Dreadful it was to see the ghastly stare,
   The stony look of horror and despair,
   Which some of these expiring victims cast
   Upon that mocking Fiend, whose veil, now raised,
   Showed them, as in death's agony they gazed,
   Not the long-promised light, the brow whose beaming
   Was to come forth, all conquering, all redeeming,
   But features horribler than Hell e'er traced
   On its own brood; no Demon of the Waste,
   No church-yard Ghole, caught lingering in the light
   Of the blessed sun, e'er blasted human sight
   With lineaments so foul, so fierce as those
   The Impostor now, in grinning mockery, shows:
   ' There, ye wise Saints, behold your Light, your Star,
   Ye *would* be dupes and victims, and ye *are*.
   Is it enough ?   Or must I, while a thrill
   Lives in your sapient bosoms, cheat you still ? ' "

Thus it is, pursued to the very grave-edge, these victims loathe the drug they once loved. Business is gone, family broken, friends lost, moral sense blunted or destroyed, mind incapable of healthy action, body wrecked, and they see no hope here or hereafter.

They will lie and steal, do almost anything to obtain the drug with which and without which they are finally in a veritable hell. The face becomes sallow and soggy, the eyes bleared and expressionless, and the final result is either death or insanity. Some persons go on using these drugs for years before the symptoms here described supervene ; some are thus affected in a few months.

Having reached this stage they cannot arouse themselves from their terrible infatuation. Gloomy and hopeless, the world and the people in it no longer interest them. A patient whom I saw some years ago, a young Spaniard, was suffering from insanity from the use of morphine. Hour by hour he would sit folding, refolding and cutting paper into small bits with an old lancet. He would speak to no one, notice no one. This went on for months. He was sent to an insane asylum finally, and on searching his trunk some two score morphine bottles, as well as the greater part of the trunk, was filled with these minute scraps of paper. I have never been able to learn what became of him.

A lady patient of mine, well advanced in years, would save scraps of tin, old bits of rags, glass stoppers of bottles, and the like, setting great store by them. She laughed heartily at her collection, and threw them away one week after her emancipation.

Sometimes there is a mawkish sentimentality exhibited toward the opposite sex ; sometimes there is mock modesty or direct abhorrence.

In the majority of instances these people are great liars, especially about matters concerning their habituation ; often also about trivial things, where falsification is absurd and absolutely without excuse.

Occasionally there is a loss of connection between ideas in talking, incoherence and silliness. The speech is, as a rule, slow and somewhat drawling, and often interrupted to wet the lips, which become dry and parched. Severe pain in the head and about the region of the heart is sometimes complained of, also in the "small of the back."

In some cases, more especially those of an intensely nervous organization, the prolonged abuse of opium or morphine produces a condition characterized by cerebral excitement, analogous to that of delirium potatorum. These people are, however, less violent, and the affection usually passes away in a short time, without treatment.

THE NERVOUS SYSTEM.

On the nervous system the effects of opium and morphine are most manifest. Taken at first to relieve pain and disorders of this system, having their chief action upon it at all times, their continued use reacts with deadly intensity. Twitching of isolated muscles, trembling of the hands and of the tongue, when protruded, and occasionally paralysis of one eyelid, are seen. The pains that supervene have no distinctive character, as do those that come from the abuse of chloral.

Itching of the whole or parts of the body is sometimes very troublesome. Herpes zoster (shingles), an eruption following the distribution of the nerve filaments upon the trunk, occasionally occurs, as also does urticaria (nettle rash). The very disorder for which the drug was first taken, is, in some instances, aggravated or perpetuated.

This was first noticed by Dr. T. Clifford Allbutt*, of England, and I have seen the same in some instances.

In the condition of delirium sometimes occurring, the pupils, usually contracted, are occasionally irregular and dilated.

The whole nervous system is unstrung, or more properly, too highly strung, so that it vibrates to little things that in health would pass unnoticed. They jump at the falling of a book or the shutting of a door.

Hysterical women still continue to have their customary attacks, sometimes in an aggravated form.

The following interesting case history has been kindly sent me by Dr. Judson B. Andrews:—

A woman,† thirty years of age, single, seamstress, with no hereditary tendency to insanity; was of a highly nervous and excitable organization, emotional and irregular in feeling; at times buoyant and lively, and then gloomy and depressed. Her health during early life was delicate, though she suffered from no definite form of disease. At the age of twenty, in April, 1862, she was seized with pain in the head. It was of short duration, but very severe, and during its continuance the patient was delirious. Attacks of the same character, both in the severity of the pain and the mental disturbance, have occurred since, at intervals of from one to three months.‡ In 1864 she had acute rheumatism, and in 1865 a severe attack of diphtheria.

After the local disease of the throat had apparently subsided vomiting supervened, and was repeated every few hours for some five weeks. To relieve this condition and

* *Practitioner*, 1871.

† *American Journal of Insanity*, July, 1872.

‡ From the subsequent history of the patient, especially while in the asylum, we are led to believe that these attacks of delirium took place at menstrual periods.

procure sleep, hypodermic injections of morphia were successfully employed for about one week, and the patient
rapidly regained her health. Some two years after this, or
in July, 1867, she had an attack of inflammation of the
bowels and peritoneum, and for four weeks was delirious
most of the time. She improved somewhat in health, but
for the four months succeeding had frequent attacks of
frenzy, during which she often threatened to take her own
and her mother's life, and became very difficult to control.
In October following she had improved so far as to pass
from the immediate charge of her physician. Soon after
this he ascertained she was using hypodermic injections of
morphia, to relieve pain in her limbs and different parts of
her body. I quote from his letter:—

I was informed that she was using it (morphia) to a considerable extent, and called immediately to explain to her the effects
and danger attending the practice. I believe every effort was
made that could be to prevail upon her to desist, but all to no
purpose. She was cunning and artful, and would almost always study out some plan to get the morphia. She has used
as much as two drachms in a week, in one or two well-authenticated instances. The usual amount was one drachm per
week. She used but little, if any, for three or four months
before she was sent to the asylum, for it was very difficult for
her to get it. She has acted very strangely ever since her first
sickness. She has been truly a mystery, which no one could
solve.

Her mother says:—

That for years she has complained of pain, and pressed her
hand on either side of her head, with the exclamation, "Oh,
mother, mother, I shall die!" That for six years she has complained of such soreness of the head that when she passed her
hand over it, in smoothing her daughter's hair, she would cry
out: "Oh, mother, don't; it hurts me so!" That five years

ago, in 1867, she was obliged to call in help, as the patient
threatened and intended to take her own life. That both before and after she began the use of morphia, her conduct was
peculiar and erratic; that she was emotional and easily disturbed by trifles. That after the morphia habit was known,
her conduct for many years preceding was wrongly attributed
to this cause.

A few weeks before she was sent to the asylum she
passed into an acutely maniacal condition, in which she
was sleepless, ate little and irregularly, lost flesh and
strength rapidly, and became quite feeble. She was de-
structive of clothing, pulled her hair out, was noisy, incoherent and violent; opposed care, wandered about, and
was with difficulty controlled. In this condition she was
admitted to the institution, on the fifth of May, 1871.
She was carried to the ward and placed in bed. Examination revealed scars and ecchymosed spots, covering nearly
the whole of the body which could be reached by her own
hand. She asserted that she had employed the hypodermic
injections for three and one-half years, once, and much of
the time twice, a day, making in all about two thousand
injections; that during the last few months of its continuance she had used a drachm and one-half of morphia per
week; that she inserted the needle perpendicularly to the
surface, and often carried its full length into the tissues.
For two days she was sleepless and retained no nourishment. Chloral, in thirty-grain doses, was then administered, which was tolerated by the stomach, and secured
sleep. The vomiting gradually became less frequent and
soon ceased. She ate well, gained flesh and strength, all
maniacal symptoms subsided, and in twenty days she was
up and about the ward. Menstruation, as she said, had
been suppressed for two years. As she complained of pain

in the back and other symptoms which usually preceded it, she was placed on the use of capsules of apiol, and on the 24th of June began to menstruate, but the flow was scanty, and accompanied by much pain.   ·

During the month following she steadily gained in mental strength, and became quite stout. At time of next menstrual period the right breast swelled to an extraordinary size, so that we were obliged to suspend it with adhesive straps. It was hard and extremely sensitive to the touch. This condition of swelling and tenderness extended in a narrow ridge to the spine. The state of the breast was at first supposed to be owing to the sympathetic action of the organ with the renewed activity of the menstrual function. For two weeks applications were employed, without success, to relieve the pain and tension. At this time, on the 13th of August, the patient, in rubbing her hand over the breast, discovered an elevated point just under the skin, which, on pressure, gave a pricking sensation. This was cut down upon, and a broken needle extracted. On the 15th another needle was removed. The breast was now inflamed and extremely sensitive. August 28th, another needle was taken out. August 29th, menstruation began again. The flow was profuse, and she became at once delirious. Was talkative, restless, profane and obscene, and pulled her hair out. She continued in this condition some twelve hours, and, as she stated the next day, was entirely unconscious of what had occurred.

From this time till September 28th, from one to five needles were removed daily from the breast. Menstruation then occurred again, and was characterized as before by a similar attack of mental disturbance. After this, during the months of October and November, needles were taken from various parts of the body ; from the left breast, the abdominal parietes, the mons veneris, the labia, and vagina. Of these latter, some passed across the urethra and rendered urination difficult and painful ; others across the vagina, either end being imbedded in opposite sides. Some were removed from the thighs, from the leg, down to the ankle, from the buttocks, from about the anus, from the back as high up as between the shoulders. The largest number extracted in any one day was twelve.

On one occasion ether was administered, but the difficulty experienced in bringing her under its influence, and the mental disturbance produced by it were so great that it was not again resorted to. During the whole period, to her final illness, she retained her flesh, though she ate and slept irregularly, under use of tonics and sedatives. She was in a variable mental state, at times irritable, petulant, fault-finding, attempting to create ill-feeling between attendants, and demanding unnecessary care and waiting upon. At other times she was abnormally cheerful, gay, pleasant, and fulsome of praise of all around her.

For the first two months but comparatively little pain was felt in the extraction of the needles. The skin was thickened, harsh and dry, and almost insensible, from the prolonged and distributed use of the injections. Afterward, she suffered acutely, and often begged, with tears, that their removal might be postponed from day to day. About a month before death she had an attack of localized pneumonia, affecting the lower portion of right lung. This was accompanied by stridulous breathing, spasm of the glottis, *globus hystericus*, crying, and other hysterical manifestations. It was followed by an attack resembling muscular rheumatism, characterized by great pain and hyperæsthesia of surface. The right arm was swelled, hot

and extremely sensitive.   It was supported on a pillow and
kept bathed in anodyne lotions.   She lost appetite and
sleep, became much depressed, and gave up all hope of
recovery.   Her tongue became dry and brown, pulse rapid,
secretions offensive, and mind very feeble.   A diarrhœa
supervened and the evacuations of bowels and bladder were
involuntary.   She became unconscious, and finally coma-
tose, and died on the 25th of December, 1871.

No needles were removed during the last two weeks;
286 were taken from her body during life; 11 were found
in the tissues after death; 3 were passed from the *rectum*
during sickness; making a total of 300 needles and pieces. ·
Of this number, 246 were whole, and 54 were parts of
needles.   One was a No. 7 sewing machine needle, and
several were bent.   They varied in size from No. 4 to No.
12.   As regards position in the body, they were distributed
about as follows : in right breast 150 ; left breast, 20 ; abdomen, 60 ; genitals, 20 ; thighs and legs, 30 ; back, 20.
Of those removed after death, 5 were found in the right
and 3 in the left breast ; one in a small abscess in the epigastric, and one in the right iliac region, the point im-
pigning upon the peritoneum, which was discolored with
rust ; and one in the upper part of lower lobe of left lung.
The presence and position of the needles were indicated to
the patient by the pricking sensation occasioned by muscular movements.   They were removed in a few instances
at first, by cutting down upon them.   This proved to be a
painful, and, from the movements of the needles in the
tissues, a difficult process.   Hemorrhage from the small
vessels, at ·times, gave some trouble.   Afterwards, by
manipulation, the ends of the needles were engaged between the thumb and forefinger, and the points, forced
through the skin, were seized and the needles extracted

with forceps.   Sometimes much force was required to
withdraw them.   They changed position quite readily, and
frequently moved from one to two inches in a day.   They
produced little local irritation or trouble beyond the
pricking sensation, and did not seem to have contributed
in any notable degree toward producing the fatal result.
In regard to the presence of this large number of needles
in the system, no information could be obtained.   The patient repeatedly and persistently denied any knowledge of
having introduced them, either by the stomach or through
the skin.   Her mother, who visited the Asylum, could
throw no light upon the subject, and was entirely ignorant
of the fact until informed by us.   She, however, recalled
the circumstance that the patient purchased, at one time,
ten papers of needles, and could account for only two of
them.   They were not obtained or introduced while in
the Asylum.   She was under strict surveillance, and had
no means of obtaining any number of needles, and those
removed were all rusted and bore evidence of having been
a long time in the body.   The stomach was closely examined after death, and was in a perfectly healthy condition,
with no evidence of any previous inflammatory action.

The only theory which seems to us at all tenable, is
that they were introduced through the skin while she was
under the influence of morphia, hypodermically administered, and while suffering from hysteria.   That some were
found in positions where they could not have been inserted
by the patient, can be accounted for by their movements
in the tissues, which were observed so often during the life
of the patient.

The diseased condition of the brain and its membranes
was a cause sufficient to account for the abnormal mental
action and conduct of her who had been " truly a mystery

which no one could solve." We close this remarkable case with a transcript of the post-mortem examination.

Autopsy.—Rigor present; body well nourished; anterior surface thickly studded with small cicatrices; abdomen covered with thick layer of fat. A small abscess in abdominal wall, two inches above umbilicus, three inches by one and one-half, was filled with pus and contained one needle. A second abscess, two inches above and to the right of the symphysis pubis, immediately under Poupart's ligament, contained another needle. This pressed upon the peritoneum, which, though discolored by rust, was not inflamed. From the right breast, one whole and four broken needles, and from the left one whole and two broken needles, were removed.

Head.—Arachnoid opaque and thickened over right hemisphere. The left hemisphere was covered by a thin layer of pus, contained in the sub-arachnoid space. Marked depression of convolutions at vertex of both hemispheres. The brain substance was firmer than normal. The ventricles were empty, and the choroid plexus contained numerous small cysts upon its surface, filled with serum.

Thorax.—The lower lobe of the right lung was hepatized. A whole needle was found in the upper part of the lower lobe of the left lung.

Abdomen.—The liver was soft and fatty, and the spleen enlarged; kidneys were normal. The stomach was subjected to a critical examination. It was found normal, and there was no evidence that the needles were introduced into the system through that organ.

The vascular system, through the agency of the nerves, is profoundly affected, as already shown.

Spasm of the muscles of the bladder and rectum is sometimes present, and in some cases is distressing and lasts for a considerable time.

Symptoms resembling those produced by malaria are occasionally found. Levenstein, who was the first to call attention to this, says: "Intermittent fever, in consequence of a morbid craving for morphia, seems to be due to a certain neuropathic disposition, as it does not show itself with many patients, although they have taken large doses of the drug, and for years together. It was, however, impossible to fix on any other cause for the development of intermittent fever but the use of morphia, as the respective patients lived in regions free from malaria, and as none of the other members of the family living under the same conditions showed any similar symptoms.

"We may distinguish a light and a severe type of intermittent fever, when brought on by a morbid craving for morphia. Both forms resemble real malarial fever, inasmuch as the first paroxysms, occurring at regular intervals, seemed to disappear after the use of quinine, returning, however, very soon, although the febrifuge was continually given; that, furthermore, they were improved by change of air, but came on again from the simplest causes, such as boating, errors of diet, etc.

"The characteristic symptoms of this fever are the same as those caused by malaria: chilly feeling, up to regular shivering, headache, oppression, heat and perspiration. They differ from one another in this respect, that immediately the morphia is discontinued, the attacks disappear without any treatment, although they may have existed a long time.

"In some cases the intermittent fever sets in in an erratic manner. The patient, at irregular times, experiences an attack of fever, with chill, heat and sweating. These attacks occur from three to six times, at long intervals, not showing themselves hereafter any more at all, or only after a great lapse of time. In most cases the attacks of intermittent fever, in morbid craving for morphia, shows a tertian, rarely a quotidian, type. They are sometimes ante-, sometimes post-ponent. The attacks last from four to ten hours, and are followed by a normal condition.

"The paroxysms disappear only in exceptional cases, without the morphia being stopped. In this case the patients complain of experiencing an uncomfortable sensation, principally of an exhausting character, at the usual time of the attacks.

"The feverish attacks are accompanied by neuralgic affections of the different nerves, principally in the region of the supra-orbital, intercostal and cardiac nerves. The temperature is increased in all cases, varying from 38.5° C. to 40° C. (101.3° F. to 104° F.). The spleen is generally enlarged. The attack is followed by sediments in the urine. In the severest forms of intermittents the patients get delirious when the fever has reached its maximum, cannot be kept in bed and may become maniacal. Both forms cause great weakness and exhaustion, which last during the intervals."

I have seen but two cases that could at all be classed as such. In one there was *no* fever, and the chill came every morning at daybreak, for three mornings, missed one morning and occurred again the next. It was followed by profuse sweating, that lasted the whole day. In the other case, that of the lady who had used the drug by the mouth for sixteen years, there was fever and some intercostal neuralgia, occurring every other day, for ten days, but not followed by perspiration. A few doses of quinine broke it up. It was for this that I was first called to attend her,

her husband believing her to be suffering from malarial fever. In neither of these cases was there any enlargement of the spleen. I then discovered that she was a morphine habituè, much to her chagrin and her husband's surprise, he never having suspected it. The two following cases are those given as examples by Levenstein :—

CASE I.—INTERMITTENT FEVER IN CONSEQUENCE OF MORBID CRAVING FOR MORPHIA.

M. H., law student, 24 years old, sent to the Institution by Dr. Ewald in 1874, was suffering from acute articular rheumatism when the first injection of morphia was administered. After his recovery, although not compelled to do so through pain, he continued the injections several times in the day, increasing the doses, for the sole reason that he felt elated by them. The principal symptoms that resulted therefrom were loss of appetite, progressive emaciation, loss of strength, and increased perspiration, which frequently caused the patient to become wet all over while in a cold room and quite quiet.

Before his admission into the Maison de Santé he was troubled with feverish attacks, which came on every two or four days, at different times in the day, in the following manner: first there was a chilly feeling for half an hour, followed by heat and profuse sweating. The latter was accompanied by the general symptoms of every feverish attack, enlargement of the spleen also being present.

Present State.—Patient is a tall, muscular man ; the examination of the internal organs shows no abnormal condition, excepting an enlarged spleen. Pupils of middle size, equal, reacting well. On December 10, 1875, in the afternoon, patient received the last injection of morphia.

December 12. Patient slept in the night. In the course of the day he only feels a little sleepy. The face is red, the skin moist. Toward the evening there is nausea, pressure in the epigastrium, great restlessness, and stomach-ache. Patient moves about in bed, complains of headache, cannot get to sleep. Three relaxed motions.

December 22. Patient has had no sleep during the night, three relaxed motions, vomited once. He complains of giddiness, restlessness and palpitation of the heart. In the morning there is a chill followed by heat and profuse sweating. Vomiting, diarrhœa. Until the afternoon he felt very prostrate and exhausted. Between 3 and 5 P.M. he got up. Soon, however, the symptoms of the morning returned again, and pain in the knees, exhaustion and restlessness compel him to go to bed.

December 23. Has slept from 2 to 5 A.M. with interruptions. Profuse perspiration, nausea, intense craving for morphia, frequent paroxysms of sneezing. The sickness stopped in the course of the day. At 8 P.M. 30 grains of chloral were given.

December 24. Only three hours' rest. Feels knocked up. One relaxed motion. Much sneezing; craving for morphia. At 9 P.M. 45 grains of chloral were given, but were immediately brought up again.

December 25. Patient has had hardly any rest. One relaxed motion. Has been sneezing frequently. Emission of semen. Great prostration, even in the horizontal posture; red cheeks; craving for morphia continues for the whole of the day. Appetite small. At 10 P.M. 45 grains of chloral.

December 26. He has slept well during the night, only woke up two or three times. Pressure in the stomach, headache and palpitation of the heart come on now and then in the course of the day. At 11 P.M. 40 grains of chloral were given.

December 27. Restless sleep, much interrupted. Patient went about the room on waking up. During the day he complained of heavy pressure in the head.

December 28. Patient has slept for about three hours. Sneezing. The red color of the face of the past days was still present to-day. Although tired he could get no rest. Two relaxed motions. In the morning a warm bath with cold douche was given.

December 29. Patient has slept for nearly eight hours. Head not well yet. Severe sneezing. He feels better in himself. Toward the evening, however, an uneasy feeling came on in the legs. Three relaxed motions.

January 1, 1876. Except the sleep being restless, patient feels well.

January 3. Slept only from 3 A.M. ; ran about in a restless manner previously. Three relaxed motions. In the afternoon warm bath with cold douche.

January 13. The bodily functions are all in a normal condition. General health good. There have been no further attacks of fever.

January 14. Patient left the Institution.

He has had no relapse.

Urine.—The specific gravity varied from 1.012 to 1.020. Reduction of oxide of copper was noticed.

CASE II—INTERMITTENT FEVER IN CONSEQUENCE OF MOR-
BID CRAVING FOR MORPHIA. (IMPOTENCE. DISORDERED
SPEECH. ALBUMINURIA.)

Captain B., sent to the Institution by Staff-Surgeon Dr. Peltzer, had been using injections of morphia in consequence of severe pains from a gunshot wound in 1871.

For a time his medical attendant diminished the drug, but soon, by the advice of the latter, he purchased a syringe and bought the morphia, first at a chemist's, and afterward at a shop where they sold chemicals; he injected gradually as much as twenty-four grains per day. Several times his wife tried to stop the injections or at least to diminish the dose, but this was followed by vomiting, diarrhœa and loss of sleep, so that the doctor again recommended its further use.

The principal complaints of the patient, on account of which he, on December 20, 1875, sought admission into the 'Maison de Santé,' were : The appetite is bad, the bowels are so much constipated that they are sometimes not relieved for eight days. From time to time patient suffers from disordered micturition, having to strain rather long before the water passes. Very frequently there was congestion to the head, and during sleep quivering of the muscles of the face and extremities. Now and then he suffered from giddiness and headache. He feels unwell, principally in the morning. Impotent for three years. He was obliged to resort to alcoholic beverages as stimulants, but he was no drunkard. From September 12 until the end of October, 1874, the patient had had a shivering lasting two hours daily, followed by half an hour's heat and two or three hours' profuse perspiration. Large doses of quinine taken daily for a period of three weeks are said to have cured the fever; it is worthy of notice that the patient stopped the use of the morphia during the latter period of the feverish attacks. Taking to it again, there was the same characteristic attack every week or fortnight at first; gradually, however, the free intervals became shorter, and at the time of his admission into the Institution the intermittent had again returned to the quotidian type. A treatment with large doses of quinine for several months, resorted to by his medical attendant, proved of no avail. Patient is pretty tall; muscles and subcutaneous areolar tissue very well developed. Face red. Eyes bright. Tremor of hands, slight degree of difficulty in speaking. Patient shows great vivacity in talking; his features move quickly; his movements are brisk. The physical examination of the thoracic and abdominal organs shows no abnormal condition, except a considerable enlargement of the spleen.

The morphia was at once withheld.

December 21. Patient had a restless night, feels exhausted and knocked up; yawns, complains of cold, loss of appetite, severe headache on moving the head, and pains in the back; this is followed by nausea and at night by vomiting. Profuse perspiration.

December 22. Patient was very restless in the night; got out of bed, ran about, laid down again, perspired freely, asked for morphia. The abundant perspiration lasted till midday and was accompanied by determination of blood to the head. Patient suffered from giddiness and felt greatly tired. Appetite poor. Frequent retching, but no vomiting.

December 23. Patient has slept little. Three relaxed motions in the morning. Symptoms the same as on the previous day. New symptoms: twitchings in the extremities, excitement, sensitiveness to the light, and epigastric pains. To remove the latter symptom sinapisms to the stomach, hot poultices and cupping (four times) were attended with success. Frequent vomiting.

December 24. Patient has only slept for a few hours. A great deal of sneezing; eight relaxed motions. In the course of the day he felt well.

December 25. The pains and pressure in the region of the stomach have returned and he had also palpitation of the heart, was very much exhausted and suffered from tenesmus. Two seminal emissions.

December 26. Four relaxed motions, shivering, feels uncomfortable.

December 28. Patient has only had two hours' rest. Hands and feet burning hot. Eight motions; during the day he felt weak, complains of formication in the hands and feet.

December 29. Uncomfortable feeling continuing the whole of the day. Patient's face was of a dark red hue ; he complained of hyperæsthesia in the feet and of cold. While reading a letter from his wife he began to cry, although the contents showed no reason for his doing so. Appetite good. Two relaxed motions.

December 30. Slept from 3 to 7 A.M. Two relaxed motions. A great deal of sneezing, pressure in the epigastrium, appetite small.

January 3, 1876. Slept from 12 to 4 A.M., after running about in a restless manner. Formication in hands and feet.

January 6. General condition satisfactory. Appetite increased.

January 14. The patient has continued to recover his strength. Bodily functions normal. Sexual power has returned.

Urine.—During the first weeks of abstinence from morphia the urine contained albumen.

Patient left the Institution on January 22, in perfect health. He has not had a relapse.

Neuralgia of one-half of the face, in all respects like malarial hemicrania, I have seen in two cases. In neither were there other symptoms of malarial trouble.   Both cases yielded to quinine and arsenic.

Super-sensitiveness of the skin, sometimes of the whole body, more often of a limb, or a feeling of numbness, is not uncommon.

A fact that I had noticed before I commenced the special study of this subject is, that those opium eaters who live past middle age usually die from paralysis.   In four instances I have seen this.   In all four cases the persons took laudanum.

CHAPTER V.

ACCIDENTS INCIDENT TO THE SUBCUTANEOUS INJECTION OF MORPHIA.

There are certain dangers attending the temporary or continued use of morphia by the hypodermic syringe that deserve careful attention, in order that, if possible, they may be remedied. The first is the production of abscess and inflammation.

The majority of those who use morphia in this way are badly scarred. The skin is thickened, reddened and covered with bluish and reddish discolorations. Abscesses just forming, formed, or commencing to heal, are found here and there. Ulcers and sloughs are sometimes seen. Cysts are occasionally met with. Isolated patches of erysipelatous inflammation and gangrene are found in some instances. In the accompanying cut (p. 73) is shown the condition of the skin in a male nurse at Bellevue Hospital, who was an habitué. The photograph from which the cut was made was taken but a short time before death. I have now other patients who are quite as badly scarred. In the case of a young married lady, the skin, everywhere that the dress covered the body in front, and the limbs all over, was scarred, contracted and discolored, as though she had been badly burned and then pricked all over with India ink.

Dujardin Beaumetz (quoted by Bartholow*) relates a case where these injuries resulted in death.

These abscesses are due, in the majority of instances, to (*a*) carelessness in injecting, (*b*) unclean needles or syringe,

* " Hypodermic Medication," p. 96. Philadelphia, 1879.

(c) a dirty or over-acid solution, and (d) a low condition of the general system, predisposing to inflammation and suppuration on slight irritation.

I have never seen but one habitué who had a clear solution of morphia, and he made it up fresh each day. Abundant testimony as to the production of inflammation and abscess from the above mentioned causes can be found in my little work on " MORPHIA HYPODERMICALLY."

Those patients who exercise great care in regard to cleanliness and manner of injecting are rarely troubled with abscess. Thus one patient of mine had used morphia subcutaneously in large amount, for over six years, injecting every time, and that several times daily, into a spot upon one thigh, that could be covered by a small tea-saucer, and has never yet had either inflammation or abscess.

Indeed, some persons who exercise no care whatever to keep syringe or solution clean, are free from this troublesome complication. Such a case is related by Dr. Roberts Bartholow, as follows: " One of the most inveterate subjects I have ever encountered was a man living in the wilds of Texas, who used a glass hypodermic syringe, that had been broken many times, and mended with successive deposits of sealing wax, until only the rusty old needle remained in view, and yet he escaped all accidents."

Magendie's solution is that most commonly used. It is of the strength of sixteen grains of sulphate of morphia to the ounce of water, a few drops of acid being added to dissolve the drug. The solution made after the plan of Dr. H. M. Keyes is excellent. It will keep for a long time unchanged. He writes as follows: " Some years ago, while attached to the Roosevelt Hospital, in New York city, after repeated experiments with various tests and anti-ferments, I became

RESULT OF SUBCUTANEOUS INJECTION (see p. 71).

convinced of the practicability of making a solution of the sulphate of morphia, of the strength of Magendie's, without the aid of acid, except salicylic, and that not as a solvent, but as a preventive of decomposition, making a solution that, when properly prepared, gave perfect satisfaction after years of use, never causing abscesses, as is frequently the case when the mineral acids are used, and when carried in the pocket for months being in as perfect condition for use as when freshly prepared.

"The following directions, if followed, will give the desired result :—

Sulphate of morphia,	256 grains
Salicylic acid,	8 grains
Distilled water,	16 fluid ounces.

"Heat the water in a porcelain capsule, over a spirit lamp, until the boiling point is reached ; add the powders and stir with a glass rod, until they are dissolved. Filter through coarse filtering paper, while hot, and keep in a glass-stoppered bottle of *green* glass."

Some physicians use carbolic acid, some chloral hydrate, some benzoic acid, and some chloroform, as preservative agents. Any of these substances, present in sufficient amount to prevent decomposition and clouding of the solution with minute vegetable growths, are apt to be irritating. In my first work on morphia, a full list of the solutions used in this and other countries will be found.

Erysipelas sometimes results from the subcutaneous use of morphia.

The syringe needles are sometimes broken off in the flesh. This is, however, a rare occurrence.

The method of making an injection, undoubtedly, has something to do with the occurrence of abscess. The usual plan is to pinch up a fold of skin and pushing the needle in quickly, inject the solution slowly, *beneath* it. Some persons prefer to plunge the needle deep into the muscular tissue. It is claimed for this plan that abscess seldom occurs, and there is certainly less liability of wounding a vein.

The following is taken from my book on "Morphia Hypodermically," and illustrates very fully another danger sometimes attending the use of the drug in this way :—

Articles have, from time to time, appeared in various medical journals, at home and abroad, detailing certain alarming symptoms following immediately upon the injection, subcutaneously, of moderate doses of morphia. Such accidents have been ascribed by most authors, to the entrance of a needle into a vein, with the consequent sudden passage of the drug into the circulation ; by some, to the injection of a bubble of air into the vein ; by others, to fright attendant upon the dread of the operation and the prick of the needle ; and by still others to the rapid absorption of the remedy when a vein is not punctured. Which of these hypotheses is the correct one it is at present difficult to decide ; perhaps each may have proved a factor at certain times or in certain cases. The weight of opinion would seem to favor the idea of sudden entrance of the drug into the circulation by puncture of a vein. By a careful study of some of these cases we may be able to come to a definite conclusion.

Dr. M. E. Woodling, of North Branch, Minn., writes :— "My first case in which the hypodermic injection of morphia was tried, resulted as follows : Patient large, strong and robust-looking man. Complained of pain in the course of the sciatic nerve, and of lumbago. Injection given back of the trochanter major, patient sitting ; given slowly. I turned to lay the syringe on the table, when the patient appeared unsteady, straightened rather rapidly and persistently, and slipped from the chair, falling full length, supine, upon the floor, pale and with absent respiration. He was now perfectly limp. No response to shaking or questions. Spoke the word 'breathe,' loudly, in his ear. This he obeyed. Repeated this for about a minute, and in another he was able to sit up, but was very sleepy and unsteady, requiring assistance. In about five minutes he was able to stand, and I took him out on the street and walked around with him for an hour. I then took him home, still sleepy, but improving. The next morning he was all right, but the pain was only partially relieved. Other injections were given, with no bad consequences.''

Dr. A. Atkinson, Professor Materia Medica, College of Physicians and Surgeons, Baltimore, in reply to my fifth query, writes: ''Never had death to result, and but one accident, and that was apparent suspension of animation for about fifteen minutes, in a young lady, very anæmic, in whose case I injected one-eighth of a grain of the sulphate of morphia (the regular Magendie's solution) into the rectus muscle of the abdomen, at the repeated and urgent entreaties of the patient, to relieve an obstinate uterine neuralgia. She recovered from the neuralgia and from the effects of the morphia in three-quarters of an hour. I had, a year before, injected one-fourth of a grain into the arm of this same patient, for cardiac neuralgia, with complete relief of the pain in one hour, and with no bad effects.''

Dr. E. Jones, of Cincinnati, has kindly written me and inclosed an article* of his, bearing directly upon this sub-

* ''Some Observations on the Deep Injection of Morphia.'' *Cincinnati Lancet and Clinic*, August 10th, 1878.

ject. In it he says: "Did the needle enter an abdominal vein? Several times. The first time I became somewhat alarmed; the patient at once threw up her arms, complained of suffocation, giddiness, excessive fatigue, a severe tingling sensation following the course of the circulation. The countenance was at first livid, then flushed; the eyes became unusually brilliant; slight muscular twitchings, profuse sweating, with cold extremities, and in a few moments complete relaxation was followed by deep sleep, which lasted only four hours, when she awakened, feeling, as she expressed it, 'ever so much better.'

"The same accident occurred three times, the symptoms much milder, with the exception of a burning sensation of both eyelids of either eye and both lips, which at one time became painfully intense. The above symptoms were produced by an injection of five grains of sulphate of morphia into an abdominal vein.

"Being unable to see her for a day or two, I requested my friend, Dr. Geo. E. Walton, who had watched the case with a good deal of interest, to call and give her an injection of two grains, when she put both hands to her head and gave a cry of excruciating agony. A sharp pain darted through her head, which lasted ten or fifteen minutes; also complained of an intense itching of nose and lips, finally passing off, leaving no deleterious effects. The same accident occurred to myself, only in a less degree. These injections were also made in the abdomen."

The patient was a German woman, aged thirty-seven, and weighing about 135 lbs. The case was presumably one of fecal accumulation in the colon, with severe abdominal pains.

Dr. W. A. Neal, of Dayton, Michigan, writes in this connection as follows: "No deaths; the only accidents

were where a vein was punctured. This produced dyspnœa, great distress, and was usually followed by a chill
and the reaction by fever; but in every instance there has
been but one chill, and fever once, lasting three or four
hours.''

Edward T. Wilson, M.B., Oxon., F.R.C.P., Lond., who
has a valuable and interesting article on the subcutaneous
injection of morphine, in the St. George's Hospital Reports, for 1869, writes me as follows: ''Never either
death or accident. Nothing beyond a temporary feeling
of faintness, and on two occasions a temporary outburst of
urticaria, which soon passed away.'' I hardly think these
phenomena were due to injection into a vein; indeed, the
writer does not endeavor to account for them on this ground.
They partake more of the nature of rapid absorption
with some idiosyncrasy. Dr. E. C. Seguin makes
mention, in the New York *Medical Record*, of a lady thus
affected (urticaria) by any preparation of opium taken by
the mouth.

Arthur R. Graham, M.D., etc., of Weybridge, England,
sends me the following interesting and conclusive case :
''No deaths; but one alarming accident worth recording :
I had injected a large dose (probably three-quarters of a
grain) into the *right* forearm of a woman whom I was in
the habit of injecting daily. Almost immediately she
started up, and holding up her *left* hand and looking at it,
exclaimed, 'O, how funny my fingers feel !' and fell back
in a dead faint, with blanched lips. I immediately bound
a tape tightly around the arm, above the puncture, and
then gave brandy and asafœtida injections, but she remained unconscious, I think, for more than half an hour.
After she was sufficiently recovered to talk rationally I
loosened the tape, when she immediately fainted again.

Of course, I at once tightened the ligature and kept it so for some hours. The second swoon was less alarming than the first. In the first no pulse could be felt, nor could the heart sounds be heard, excepting with great difficulty. My impression at that time was that, had I not applied the ligature at once I should have had a fatal result to chronicle. It was the impulse of the moment to tie on the tape, and had I had time to reason I should have rejected the idea as an entirely useless one ; but in any similar emergency I should now recommend any one to try it."

Dr. E. Fletcher Ingalls*, who has devoted much attention to the hypodermic injection of morphia, reports the following case: " I have often used hypodermic injections of morphia, and always with good results, until a few weeks since, when I obtained alarming results from the administration, by this method, of one-fourth of a grain of morphia.

" The patient, in consequence of continuous watching with sick children, had become debilitated, and, as a result, suffered at times from severe pains of a neuralgic character. I was called in the night to see her in one of these attacks. The pain had commenced about twelve hours previously, and with frequent exacerbations, had steadily increased in severity until it had become unbearable.

" I dissolved one-fourth of a grain of morphia in pure water, and administered it under the integument on the outer side of the arm. Within a few seconds the breathing became stertorous, the pulse failed, the lips and countenance became livid, and the eyes were set ; respiration ceased, the radial and cardiac pulsations were lost, and the heart sounds could not be distinguished. The woman

* Chicago *Medical Journal and Examiner*, Aug. 1877.

was to all appearances dead. How long this condition continued I cannot tell; it seemed an age, but was probably only ten or fifteen seconds, for by prompt means I succeeded in resuscitating my patient.

"After a few minutes she expressed herself as much relieved. I remained with her some time, and then left careful directions with the husband in case any other unfavorable symptoms should occur. During the next few hours the patient fainted twice, but she was restored by dashes of cold water in the face."

Dr. H. L. Harrington, of Little York, Ill., refers me to the report of a case* of his, which reads as follows: "Was called a short time since to treat W. S., male, aged sixty-two, for acute dyspepsia (bilious attack) accompanied by very severe pain. Administered hypodermically, in the hypogastric region, morphiæ sulph. o.o2 gram. Before the syringe was emptied alarming syncope supervened, and occurred twice, at intervals of ten or fifteen minutes. Stimulants administered freely, artificial respiration and the use of electricity were successful in reviving the patient. Neither narcotism nor coma were in any degree present. Is it possible to attribute the syncope to the effects of the drug? Not over fifteen seconds were occupied in the operation." I think this an excellent example of the puncture of a vein with entrance of the drug directly into the circulation.

Dr. Aug. M. Tupper, of Rockport, Mass., published the following interesting case :†

"On the morning of the 22d ult. I was called to see Mr. G., who was stopping at one of our hotels. I found a healthy-looking young man, about thirty years old, suffer-

* Chicago *Med. Journal and Examiner*, April, 1879.
† Boston *Medical and Surgical Journal*, October 30th, 1879.

ing from lumbago, confined to his bed, and in considerable pain, aggravated very much by movement. Applications of mustard and an anodyne liniment were prescribed. In the evening I called again, and as the relief was slight, decided to inject some morphia directly over the seat of pain, a method I have found very efficacious in similar cases. Accordingly I injected nine drops of a solution of sulphate of morphia, one grain to a drachm of water, into a spot midway between the spine and crest of the ilium. As is my custom, the solution contained one drop of carbolic acid, which I added in order to keep it. In five minutes he expressed himself as feeling relieved, and sat up in bed to show us the improvement. I told him to lie down and keep still awhile, and he did so. We chatted pleasantly for perhaps five minutes longer, when, turning toward his wife, he said, ' I think I am going to vomit,' and turned to the side of the bed. I noticed that he looked a little pale, and before Mrs. G., could get the basin, he grew deadly pale, his eyes rolled up in his head so that only the whites were visible, the jaws were clenched, the head was drawn back, and the whole body stiffened, respiration ceasing also. I immediately went to him, dashed cold water in his face, and took the wrist to feel his pulse, which, to my horror, was not to be felt. He was in this state for perhaps a minute. I then raised him up, and looking into his eyes, which were staring wide open, saw that the pupils were widely dilated. Very soon the color began to return to his face, he was drenched with perspiration, and recovered consciousness. I laid him back on the bed, and he looked up, smiled, and said, ' I'm all right now.' The pulse was quite full at sixty, but inclined to be irregular. I gave him brandy freely, and he had no further trouble, but the pulse remained at sixty for the

6

next twenty-four hours ; he said it was usually about eighty.
I cannot verify that, for he left town the following day.

"That was certainly a very unusual effect from such a
dose, a little over one-eighth of a grain. The question
arose in my mind whether the acid could have had anything to do with it ; but I have given the same mixture a
great many times without the slightest trouble. I may add
that the solution was prepared that morning, and I injected
the same dose into the same part of the body, for neuralgia, in a female patient, that very same day, previous to
using it on this patient. I should not care to repeat this
operation on Mr. G., and advised him never to have it
done again. I would also state that it relieved his lumbago,
for the next afternoon he was dressed and down at his meals."

After seeing the report of this case I wrote Dr. Tupper,
who has courteously furnished me with the following additional facts : "My patient, I should say, was of a phleg-
matic temperament. He had taken no medicines by the
mouth before I saw him, nor while under my care. He
had never taken any narcotic in his life, he says. The
pupils were natural very soon after the effects of the dose
ceased. He writes me that his pulse remained at sixty for
a week, but that he felt first rate. Pulse rate since then has
been seventy-six. It seems to me that it must have been
a peculiar susceptibility to this drug in his case, else it
would not have had such a lasting effect upon the pulse.
I do not think it was due to the mode of administration.
He related to me afterward, that a cousin (I think that
was the relation) had very peculiar and even dangerous
symptoms from a dose of Dover's powder, some time ago,
and that the physician in attendance was detained all
night in consequence. This would go to show an idiosyncrasy in the family."

I think that the doctor is right with reference to idiosyncrasy, but I think, also, that the method of administering the drug had much to do with it.   As I have said before, where an idiosyncrasy, be it to narcotism or any other peculiar manifestation, exists, the sudden entrance of the drug into the system is certain to aggravate those symptoms; may, indeed, call forth an idiosyncrasy that the drug given by the stomach would, possibly, never have revealed.   The fact that some ten minutes elapsed between the time of the injection and the first appearance of the alarming symptoms seems to preclude the idea of the needle having entered a vein.   The symptoms are, however, exactly those that are seen when a vein is punctured. As will be seen from the results of experiments soon to be recorded, and from the conclusions arrived at by the Committee of the Medico-Chirurgical Society* of England, five minutes is abundant time for enough of the drug to be absorbed to produce its characteristic, and, therefore, its unusual effects, where idiosyncrasy exists.

Prof. H. C. Wood,† states that he has seen deep coma produced in three minutes by a hypodermic injection of morphia.   This may have been due to unusual rapidity of absorption from the cellular tissue, or to direct injection into a vein.

Here is a somewhat similar case ‡: A lady, aged twenty-four, who has been a sufferer from neuralgia every day for months, was given a hypodermic injection of one-fourth of a grain of the hydrochlorate of morphia in the subcutaneous tissue of the leg.   Alarming syncope and extreme prostration came on within five minutes after the injection was made.   The patient was not out of danger for four

* "Medico-Chirurgical Society Trans.," vol. L.
† "Materia Medica and Therapeutics," Phila., 1877, p. 205.
‡ "Medico-Chirurgical Transactions," vol. L.

hours after, and was too ill to leave her bed for two weeks. The neuralgia did not return for some months.

Dr. Francis H. Miller, of East New York, formerly House Surgeon of St. Peter's Hospital, Brooklyn, writes me: "Several times, when I suppose my fluid entered some small vein, the patients complained of sudden weakness, faintness and dizziness, and almost syncope."

Dr. Samuel W. Francis, of Newport, R. I., writes me: "I have heard of two or three cases where extreme syncope set in, the patients being restored only with great difficulty."

Dr. J. S. Jewell, Professor of Nervous and Mental Diseases, Chicago Medical College, writes me: "I have never had any serious consequences follow morphia injection, but have seen temporary unpleasant symptoms (vasomotor disturbances), vertigo, mental confusion, etc., a few times."

Dr. Geo. R. Fowler,* of Brooklyn, states that he has twice had alarming symptoms from puncturing a vein. He believes that this may be avoided by making the skin of the part to be punctured tense, and introducing the needle at a right angle to the axis of the limb.

Prof Nussbaum,† of Munich, has published an interesting account of an accident that happened to himself. He had made use of the hypodermic method of giving morphine to himself, as often as 2000 times, using sometimes as much as five grains of morphia at an injection. One day he accidentally injected two grains of the acetate of morphia into a vein, and did not recover from the dangerous effects for two hours. He has seen the same symptoms, in a less degree, in two of his patients. He advises

* N. Y. *Medical Record,* Aug. 15th, 1874.
† *Medical Times and Gazette,* Sept. 23d, 1865.

slow injecting, and withdrawal of the piston if a vein is punctured.

The following interesting remarks are clipped from an English journal,* and bear directly on the question in hand :—

" ' Observer ' remarks that ' Spectator,' in the *Journal,* of April 12th, very accurately described what always happened when a vein has been pierced and morphia injected into it, although he might have added (as no doubt it occurred) that the person injected also experienced a strong taste of morphia ; and probably, also, an unusually. large quantity of blood flowed from the puncture. It is a very serious accident to inject morphia into a vein, but it need never happen, if the operator, thrusting the instrument under the skin, will draw up the piston, when, if the point be in a vein, blood will be drawn into the syringe. ' Observer ' knows a gentleman who for years has been in the habit of injecting himself with morphia, three and four times a day, to whom the accident has frequently occurred, accompanied by the symptoms described by ' Spectator.' He is much alarmed at the time, and is afterward careful to draw up his piston, but in three or four months he begins to be less cautious, until he gets another reminder. As to the necessity of drawing up the piston, there cannot be two opinions ; for besides the symptoms certain to follow the introduction of morphia directly into a vein, there is the danger of air entering as well, should care not be taken to prevent it. ' Observer has known an habitual morphia taker by injection to contract albuminuria. The albumen would greatly diminish on the daily quantity of the injection being lessened, and entirely disappear in forty-eight hours, when the morphia was wholly discontinued.

* *British Medical Journal,* May, 1879.

" 'Injector' was, until very lately, for nearly two years, one of the victims of morphia, and during that time he five times thoroughly, and twice partially, experienced more or less of the horrible symptoms sketched by 'Spectator' in the *Journal* of April 12th. The sensations were as follows: 1. A dull, gnawing pain in some decayed teeth, accompanied by a metallic taste in the mouth. 2. A pricking and tingling of the forehead and cheeks, somewhat like prickly heat; but this soon increased, spreading to the ears, neck, arms and chest (but not below the waist, although the morphia was injected into the calf of the leg). This pain soon became almost unbearable, but it was entirely eclipsed by what 'Spectator' calls 'throbbing,' but which 'Injector' says would be better represented by imagining twenty blacksmiths confined in his head, with each an India-rubber-headed sledge hammer, and each trying to make the best of his way out. Imagine, at the same time, that you are suffering from the first mentioned broiled feeling; that your skin feels as if about to burst; your eyes as if already started from their sockets; your lips as if they did not belong to you—then you may have a faint idea of what 'Spectator' wishes to describe, and what I, who have five times felt it, yet feel powerless to lay before you as I ought." He had always thought that this condition was caused by: 1, too rapid injection; 2, too much at once; 3, the solution being too strong; and 4, by its being injected directly into a tolerably large vein. When the symptoms have occurred, he has always noticed that injection has taken place at one of two spots—probably into the same vein each time in each leg—one on each leg. Again, it only occurred when he had to inject a large quantity in a short time; and he always used a very strong solution of acetate of morphia (forty grains to the half-ounce of water). He advises that when one feels any of the symptoms coming, he should walk about as rapidly as he can. He has relieved his worst attacks in this way, and has warded off others by violent exercise in his room, as soon as he felt the pain in his teeth, or the metallic taste or the "prickly heat." He thinks that corroboration of his belief as to the cause of the symptoms is afforded by Dr. Pepper's description of the results of injecting milk into the veins of anæmic patients.

The case referred to by "Observer" reads as follows :* "Scarcely has the fluid left the syringe when the most intense feeling of irritation and pricking is felt in the skin, spreading from the puncture rapidly all over the body. At the same time the skin becomes suffused with a bright blush. The heart's action then becomes greatly quickened, and there is a throbbing, rushing feeling through the head. The hands are somewhat swollen and the lips get a glazed appearance. In one case that I had, the patient became suddenly unconscious, as if knocked down by the sudden shock; and in all the cases where these symptoms have appeared the general disturbance has been very great and the attack of a severe character. The symptoms generally subside gradually, leaving behind great pain in the head.

This gentleman gives these as the main symptoms of several such accidents that have occurred in his · practice, and characterizes their occurrence as something novel and important.

Dr. J. A. Houtz,† of Logansville, Pa., who is a staunch advocate of the hypodermic method, says : "The greatest danger is in injecting into a vein sufficiently large to carry the whole dose at once into the circulation. That can be

---

* *British Medical Journal,* April 12th, 1879.
† Philadelphia *Med. and Surg. Reporter,* Oct. 18th, 1879.

avoided by selecting a place where the large veins are least numerous, and by injecting, say a third of a dose, and then waiting eight or ten seconds, when, if in a vein, the symptoms will show themselves. The first symptoms are a feeling of great fullness of the head and intense flushing of the face, coming on within a few seconds after the operation. Such, at least, was the case in a patient of mine.''

A case of syncope and prostration is reported by Dr. E. Wenger, of Gilman, Ill. The amount of morphia used is not stated.

F. Woodhouse Braine, F.R.C.S., etc., publishes the following case:* '' Mrs. H. C., aged thirty-five, in good health otherwise, had been kept awake seventy-two hours by intense neuralgic pain on left side of head, face and neck, arising from a carious molar tooth on the left side of lower jaw. She was injected with one-third of a grain of acetate of morphia. At 1 A.M., on June 28th last, the morphia, dissolved in about four drops of water, was introduced under the skin of the left arm, just over the insertion of the deltoid. No blood appeared at the puncture. In about fifteen seconds tightness of the chest and difficulty in breathing was complained of, and the patient asked to be raised, saying she felt as if she was dying. Her face and lips now became pale ; speech became indistinct (not inaudible) ; pulse irregular ; some spasm of the facial muscles took place, and she fell, to all appearance, dead. Cold water was freely dashed over her face and chest, and as she was unable to swallow, her tongue was rubbed over with sal volatile, and ammonia applied to her nose, artificial respiration being kept up at the same time. During this time her face was blanched, pulse not to be felt, and

* *Med. Times and Gazette*, Jan. 4th, 1868, p. 8.

respiration not to be perceived. Insensibility continued for about three minutes; then, happily, one or two feeble beats of the pulse, and a shallow inspiration or two, showed returning animation. She then became conscious; pulse feeble but regular; respiration slow; fingers remained numb and both thumbs were firmly drawn into the palms of the hands. This passed off in about six minutes, leaving her feeling very ill, but free from the neuralgic pain, which did not return. There was no feeling of nausea, and no attempt at vomiting during any part of the time."

Mr. Arthur Roberts* publishes the following cases :—

"Sir:—The case described by Mr. Braine, in your last week's journal, of an unusual effect of subcutaneous injection, is what I have seen in two instances, but nothing like to such an alarming extent. One was in a gentleman whom I had injected several times previously, the other in a lady. I have also partly noticed it when I have injected myself. In the first case, a few minutes after the operation, the face became intensely flushed; this was followed by vomiting, and then a dead faint and struggling for breath, the pulse scarcely perceptible. These cases, and the effect on myself, taught me, when injecting a patient for the first time, never to give more than the sixth of a grain—wait a quarter of an hour longer, and then give the remainder of the dose, after ascertaining how the first injection was taking effect. Women, I have found, are generally bad subjects for subcutaneous treatment; for they get frightened and nervous—in fact, one woman told me that though the morphia taken by the skin did her more good than by the mouth, yet she preferred the latter, for the instrument frightened her. I have used my needle over three hundred times, and I have always noticed one

* *Medical Times and Gazette,* Jan. 11th, 1868, p. 53.

fact, that if the wound bleeds after the operation the morphine enters the system much more powerfully and rapidly ; and I always know when it is going to bleed, by the operation giving a good deal of pain. When this is the case I withdraw the instrument, to see if the puncture bleeds ; if it does, I try a fresh place."

Bartholow,* in his useful and able little work, calls attention to this danger in these words : " In practicing the hypodermic injection it is important to avoid puncturing a vein. Serious depression of the powers of life, fainting and sudden and profound narcotism have been produced by injecting a solution of morphia directly into a vein. Fatal collapse might be induced by injecting air into a large vein, along with the solution."

Syncope, etc., as we have seen, from this cause, is common ; narcotism rare, the drug seeming to exhaust itself in its initial action, or to produce a condition of the system in which narcotism is wholly or partially impossible.

Dr. Corona† (*Giornale di Medicina Militaire*), summing up the results of his experiments on animals, says : " The injection of the two poisons (morphia and atropia) *into the veins* showed that a much smaller dose was sufficient to produce rapid and grave poisoning, but even then the morphia produced its action instantaneously, and its action always superseded that of atropia."

J. Pennock Sleightholme, L.R.C.P., Lond.,‡ reports the following case : " A young man of sound constitution and good health, who had never before taken morphia hypodermically, partly as an experiment and partly with the hope of relieving some slight restlessness, injected himself, at about 3 A.M., with one-sixth of a grain of morphia.

* " The Hypodermic Method," Philadelphia, 1879, p. 32.
† Edin. *Med. Journal*, Dec. 1876. *Practitioner*, 1877, p. 132.
‡ *Practitioner*, July, 1871, p. 25.

Immediately after the injection he fell down on the floor in a state of syncope, and had slight convulsive movements on one side of the body; consciousness did not entirely leave him, and after lying still for about ten minutes he was sufficiently recovered to be able to go to the next room and help himself to a couple of glasses of sherry. After this the feeling of faintness gradually passed off, and he slept for about two hours, but on rising, at eight o'clock in the morning, the same feeling of faintness returned, accompanied with great pallor. These symptoms were relieved by a dose of brandy, but did not entirely cease until noon the same day."

A physician in the South, who is a slave to this habit, writes me : " Several times I have been unfortunate enough to puncture a vein, and to introduce some of the solution directly into it. Immediately I feel a peculiar tingling all over me, from the tips of my fingers to the ends of my toes. The skin of my head feels as if a myriad of pins were penetrating it. This feeling passes off in from three to five seconds. Sometimes it has been followed by a turgescence of the vessels of the brain, causing a great fullness and throbbing, with slight headache following."

In a most interesting and instructive letter from Dr. Wm. W. Cable, of Pittsburg, Pa., I find the following : " I have spoken of minor accidents which sometimes occurred. In all that I have seen they were caused by the injection of the morphia directly into a vein. A series of phenomena instantly take place. The patient describes the first sensations as the stinging of bees all over the body, with difficulty of respiration, and intense congestion and swelling of the face and body. In one case that I saw the face was so swollen that in five minutes all traces of the natural features were lost. This condition of affairs calls for prompt action. If possible, the patient must be kept in motion and applications of cold water be made to the face and spine. If the patient falls the limbs must be raised, and all methods used to keep the heart acting, for if you can *bridge over* ten or twelve minutes the patient is safe. To prevent being ' *struck*,' as he calls it, one patient of mine carries a cord, which he throws over the arm, and if an unfavorable symptom occurs, he uses it as a tourniquet, and in a moment the result is apparent in the extravasation of the blood and the morphia from the wounded vessel. This is a safe condition, as afterward no rapid absorption can occur.''

This is a companion case to that reported by Graham, where ligation of the limb proved to be of great practical importance.

An interesting series of experiments, bearing directly upon the use of the ligature in such cases, were made by Mr. Georges,* at the Paris Society of Practical Medicine, some of which were conducted for this gentleman by M. Claude Bernard. These experiments consisted in ''injecting poisonous substances into the cellular tissue, with the view of showing the far greater safety and certainty of the hypodermic method as a means of administering, medicinally, highly-poisonous substances. He injected quantities of codeine, atropine and especially strychnine, which would surely cause death in the absence of precautions for preventing the too rapid introduction of the poisons. These injections were practiced without danger in the dog's paw, the passage of the poison into the veins being checked by the forcible application of a ligature around the paw. To render the experiment still more striking, he resolved to employ injections of the most dan-

* *Medical Times and Gazette*, June 14th, 1865, p. 42.

gerous of poisons—curare—and M. Claude Bernard conducted them for him. A solution containing about five centigrams of curare (sufficient to kill more than fifty dogs of the size of the one operated upon), was injected into the paw, and in twenty minutes the animal fell on its side. The paw was now firmly tied, and at the end of about twenty minutes the animal arose. *Whenever the ligature was loosened he again fell down, sometimes at the end of ten minutes, and sometimes in a shorter period,** and in this way it became possible to dose with complete certainty, according to the effect desired to be produced, the quantity of poison to be absorbed. The next day the dog was found on his three paws, only suffering from the swelling caused in the fourth by the injections. The ligature was removed and he was soon all right.

" The same experiment performed on another dog was followed by the same results, the animal being caused to fall or rise at the end of five, ten, or fifteen minutes, accordingly as the paw was tied or untied. This dog, however, next morning, on the removal of the ligature, fell down again, all the poison not having had time to become eliminated by the urine, so that it was necessary to reapply the ligature. M. Georges points out the superiority of the endermic method, when we have to administer powerful substances, as we may apportion the dose with an exactitude, according to the tolerance of the disease and idiosyncrasy of the patient, quite unattainable when administered internally."

Dr. Alonzo Clark, Professor of Theory and Practice of Medicine and Clinical Medicine in the College of Physicians and Surgeons, New York, kindly gives me the details of the following case: "There was brought into Bellevue

* Italics mine.

Hospital, some years ago, during his term of service, a young woman, aged about twenty-five, suffering from trismus. The jaws were so firmly locked that it was necessary to break out a tooth, in order to administer food and medicine. All ordinary medicine failing, on the evening of the second day the house physician determined to treat the case with hypodermic injections of morphia. He gave three injections of fifteen minims of Magendie's solution, with two hours' interval between the doses, and finding that no effects of the morphia were apparent at 2 A.M., he gave an injection of twenty minims. When he returned to the ward, at 4 A.M., the patient was dead. The nurse, on being questioned, stated that the patient was "asleep" before the doctor left the ward. The arm in which the puncture had been made was examined by Dr. Clark and others, and *over the point of the last puncture a little discoloration, as from extravasated blood, was apparent, and which, on careful dissection, was found to mark the track of the needle, which had opened directly into a vein.*

In this case the patient probably died almost immediately, the action of the morphia being shock-like, and its effect the more intense as one grain and a half had already entered the system by the skin.

Prof. Wm. T. Lusk, of this city, writes me of a case of syncope following immediately upon the injection. No blood appeared at the point of puncture.

Another case of death from injection into a vein is reported to me by Professor Willard Parker. An injection of morphia, to relieve the severe pain of neuralgia, was made into the temporal region of an apparently healthy young man. Death was almost immediate. The case was in the hands of a physician in Connecticut. I have repeatedly written, asking for full particulars, but cannot get them. It is a strange fact that all the cases where injection into the temporal or infra-orbital region is mentioned by correspondents and by some authors, have been attended by either intense narcotism or death. This is, of course, not a uniform occurrence, but it has happened sufficiently often to call our attention to it, and urge caution in its use in this situation.

In this connection, and in point of history, the following quaintly worded extract from a diary, which appears in *Pepy's Journal*,* of May 16th, 1664, is of interest. "With Mr. Pierce, the surgeon, to see the experiment of killing a dog by letting opium into its hind leg. He and Dr. Clark did fail mightily in hitting the vein, and in effect did not do the business after many trials; but with the little they got in, the dog did presently fall asleep and so lay till we cut him up."

The different effects produced by the same accident on different persons seems to be due to some peculiarity of constitution with reference to morphine. The drug is thrown so rapidly into the circulation that it carries everything before it, seeming to instantly overwhelm the vital powers. The brunt of its action seems to be exerted on the heart, and the key to proper treatment is thereby afforded. To whatever cause due, the effect is essentially the same, the difference being only one of degree. That syncope and vaso-motor disturbances are ever due to the injection of air into a vein with the solution, I very much doubt. In the first place the morphia itself is quite sufficient to produce the symptoms, and in the second place not more than a bubble of air is ever left, by carelessness, in a syringe, and this is not sufficient to produce these symptoms. To settle this matter, I purposely injected into

* E. P. Wilson, "St. George's Hospital Reports," 1869.

the median cephalic vein of my own arm twice as much
air as this, with a solution of warm water, and without any
bad effects; in fact, no effect whatever. My arm, above
the point of puncture, was protected by a ligature that
could have been drawn tight at a moment's notice, had
there been any untoward symptoms. The following day I
injected one-sixteenth of a grain of the sulphate of morphia
into another vein, with the effect of producing sudden
dizziness, a feeling as if the head would burst, pricking
and tingling of the nose, suffusion of the face and eyeballs,
dilatation of the pupils, faintness and nausea. The pulse
was first greatly accelerated, and then fell to about 65, and
remained so all that day. My pulse in health is 74. This
was done at 10.30 A.M., and I did not fully recover until
about 3 P.M. It may be well to state that morphia, either
by the mouth or skin, always has a very unpleasant effect
on me, while opium has not. Instead of causing sleep
and soothing irritation, the former makes me nervous,
"twitchy" and somewhat light-headed. In both cases the
needle entered the vein, as it was made prominent by the
ligature, and blood appeared at the point of puncture. As
soon as the injection was made the ligature was loosened, I
being ready to pull it tight at a moment's notice.

M. Calvet* presents a physiological research of the action
of morphine upon the various functions of the organism.
A clinical study of morphine as a therapeutical agent,
especially in the relations of acute to chronic morphinism.
In the first he observes that both intravenous as well as
subcutaneous injection of the hydrochlorate of morphine
accelerates respiratory movements, succeeded by a period
of retardation, and produces sometimes a momentary arrest
or respiratory syncope. The same relative effects occur

* " Thèse de Paris," *N. Y. Med. Journal*, Sept. 1877.

with the cardiac movements; at first accelerated, followed by retarded pulsations; sometimes even by cardiac syncope. During this time animal heat exhibits analogous phenomena, namely, the elevated is followed by lowered temperature. In fact, the absorption of morphine, whether by intravenous or subcutaneous injection, produces a very marked influence upon the reflex actions. In certain cases the period of exaltation does not occur, but the temperature becomes lowered, and the respiratory and cardiac movements are slower. Though he has not finally completed his researches, M. Calvet advances the opinion "that the above phenomena are the dyspnœa, dizziness, etc., sometimes seen during the operation of intravenous injection of milk." A study of these cases* (milk injections) shows us that these phenomena rarely if ever present themselves until a large bulk of fluid (from two to six ounces) has been added to the blood, and that, therefore, the argument does not hold good.

An interesting and novel series of careful experiments on man and animals, made by Dr. Gaspar Griswold,† then house physician in Bellevue Hospital, this city, although not made with that end in view, seem to demonstrate very clearly that a powerful and possibly irritant medicine (aq. ammonia dil.) when injected into a vein in quantity does not produce any untoward symptoms, but, on the contrary, were always found to have the happiest, and sometimes a most marvelous effect.

Another question arises here: Is it possible for such phenomena as have been ascribed to the injection of morphia into a vein to take place without such puncture?

* Pepper, *N. Y. Medical Record*, Nov. 16th, 1878. H. H. Smith, *Ibid.* Joseph H. Howe, *Ibid.*, Dec. 7th, 1878, and Jan. 4th, 1879. J. S. Prout, *Ibid*, May 11th, 1878.

† " The Intravenous Injection of Ammonia," *N. Y. Medical Record*, June, 1879.

7

I think so. Some of the cases reported would seem to prove it; notably that by Dr. Tupper, where the symptoms did not appear until ten minutes after the injection was made. Every insertion of a hypodermic needle, of necessity, cuts across or opens a number of capillary vessels, to which is undoubtedly due, to a certain extent, the rapidity of absorption when drugs are given in this manner. The Committee of the Medico-Chirurgical Society* came to the conclusion, from experiments on men and animals, that absorption of a sufficient amount of the remedy to produce decided symptoms took place in from four to ten minutes.

"Experiments on a healthy man, aged thirty-two. One-sixth of a grain of acetate of morphia was employed :—

Symptoms.	Skin.	Mouth.
Absorption . . .	5 min.	110 min.
Pulse increased . . .	8 beats.	None.
Pulse lowered . .	12 "	10 beats.
Headache . .	36 hours.	10 hours.
Nausea	46 "	3 "
Pulse, its nat. standard .	22 "	8 "
Incapacity to work .	7 "	None.
Total duration of symptoms	46 "	11 hours.

I have found that some drugs, notably jaborandi, when used hypodermically, manifest their peculiar symptoms in so short a time as one minute, no vein of any size being punctured. It is to be supposed that one drug is absorbed from the subcutaneous cellular tissue with about the same rapidity as another, but that each drug *manifests its presence* in the circulation with a difference in point of time, according to its peculiar action, or to some idiosyncrasy of the patient. Thus, while morphia may be absorbed as rapidly as pilocarpine or jaborandi, it does not, save in

* " Medico-Chirurg. Transactions," vol. L. p. 570.

certain persons or at certain times, give evidence of its presence by recognizable symptoms so early or so decidedly as the latter, *unless the patient manifests some idiosyncrasy.* From this it would appear possible, in certain cases, for a very rapid absorption to take place, and sudden over-whelming of the heart by the drug to occur without the puncture of a vein. As the needle does not go deeper than the subcutaneous cellular tissues, in a large majority of the cases the immediate treatment would be as effective in one case as in the other.

At a recent meeting of the New York Pathological Society,* Dr. Amidon presented some microscopic speci-mens exemplifying the pathology of hypodermic medi-cation. He said that although hypodermic medica-tion had been in vogue twenty-five years, or according to the claims of some, forty years, he was not aware of any accurate investigation of the relations between the hypodermically injected mass and the skin. He had injected Prussian blue (a weak solution) into the skin of moribund subjects, and a portion of skin was excised after death. The hypodermic injection was given in what he considered the best manner, namely, the pinching up a fold of skin and introducing the needle horizontally. The hypodermic injection was found to occupy a space three and a half centimetres in diameter and one milli-metre in thickness, tapering in shape. The location of the hypodermic injection varied according to the amount of adipose tissues in the subject. In those who had but little adipose tissue the hypodermic injection remained immediately below the skin, while in those who had much, the injection diffused itself. It would be seen in the specimens presented that the hypodermically injected mass

* New York *Medical Gazette*, Jan. 10th, 1880.

lay close to the arteries and veins; sometimes it completely surrounded an artery or vein. This, together with some other experiments, proved that it was by the blood vessels and not by the lymphatics that absorption took place.

In order to still further test the matter, he had injected muriate of pilocarpine into the ankle and into the supraclavicular region. The physiological effects of the drug (diaphoresis, salivation, etc.) were produced in both cases in about the same time, varying in different subjects from one and a half to four minutes; there was no appreciable difference. If the absorption had been by means of the lymphatics, the injection in the supraclavicular region would have produced its effects much more rapidly than that injected into the ankle. In one of the slides the section had fortunately been in the line of the puncture of the needle, and showed that considerable injury had been done to the tissues.

I am firmly convinced that no physician should be held free from blame in case of accident where he has not had a ligature or tape loosely encircling the arm above the point of puncture. At the first intimation of danger this should be pulled tight and kept so for several hours, being loosened gradually, thus permitting but a gradual entrance of the drug into the general circulation. With

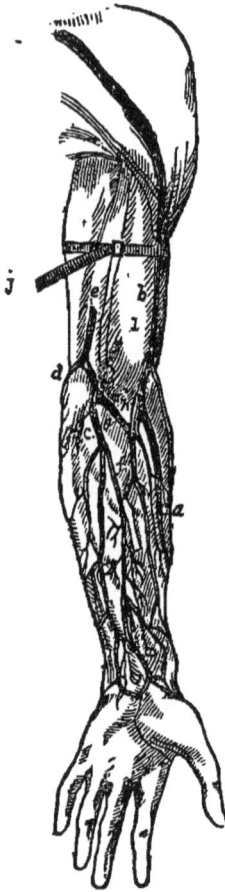

(*Agnew.*)

a, Commencement of the Cephalic Vein; *b*, Main Trunk of the Cephalic Vein; *c*, Anterior Branch of Basilic Vein; *d*, Posterior Branch of Basilic Vein; *e*, Basilic Vein; *f*, Median Vein; *g*, Median Basilic Vein; *h*, Median Cephalic Vein; *i*, Biceps Muscle; *j*, Tourniquet.

this precaution it will be seldom necessary to treat such alarming symptoms as are here recorded. A tourniquet for this purpose is here shown. It consists of a strap or heavy tape, at one end of which is sewed a "patent buckle," that will catch and hold at any point. In the absence of this or a skate strap with such a buckle, any cord or tape, so arranged as to be pulled tight at a moment's warning, may be used. The treatment of such condition when already established, is summed up in one word—*stimulation.* Whiskey and ammonia hypodermically, cold affusion, electricity, when there is a battery at hand, and hot bottles to the præcordia.

In some instances tetanus has followed the use of rusty needles, in one case resulting in the death of the victim, an habituè.*

At Southsea, recently, an inquest was held upon the body of Mrs. Frampton, wife of a lieutenant and adjutant in the Royal Marine Light Infantry.

The husband of deceased deposed that his wife was twenty-five years old, and that in 1871, previous to giving birth to a child, she suffered greatly from sickness. A surgeon, to alleviate this, used morphia by the hypodermic method, always injecting the solution himself. Some time since deceased assured him she had entirely given up the use of morphia.

On the previous Thursday he found the deceased suffering from convulsions. She grew worse and died the following morning. Since her death several bottles had been found in her wardrobe, tied up in a parcel and secreted, together with five or six small cases, each containing a hypodermic syringe.

* "The Hypodermic Injection of Morphia ; its History, Advantages and Dangers." N. Y., 1880.

Mr. Cruise, pharmaceutical chemist, said that at first he refused to serve the solution, but on reference to his junior assistant he was informed that Mrs. Frampton had been frequently supplied with the solution. In August, September and October, he supplied nine bottles each month, the last being on the 30th ult.

Dr. Norman described the state he found deceased in, and stated that when Mr. Norman and Dr. Jackson were called in they discovered on the upper part of both arms a large number of old scars, which they were informed were the result of hypodermic injections five years ago. There were no recent marks about the arms, but upon both thighs there were a large number of similar marks, and also several marks of recent punctures. Around some of these latter there was a redness of the skin in different stages, and one particularly had the appearance of having been made within twenty-four hours. He was of opinion that Mrs. Frampton died from tetanus, caused by the punctures made in the thighs for the purpose of injecting a solution of morphia. He had been shown three syringes, all of which were in a dirty condition, apparently not having been wiped dry after using. The steel needles were in a very rusty state, which would be likely to set up inflammation.

The jury returned a verdict to the effect, " That the deceased died from tetanus, or lockjaw, caused by inflammation arising from punctures made by the deceased herself, for the purpose of subcutaneous injection of a solution of morphia."*

The newspaper report of the following case has been kindly sent me by F. W. Barkitt, L.R.C.P., etc., Dublin, Ireland. The extract is from the *Irish Times* newspaper, October 23d, 1870, giving an account of the coroner's inquest :—

* *Medical Press and Circular*, Nov. 29th, 1876.

The patient was a governess, single, aged 56, and addicted to the morphine habit, using the drug hypodermically.

"Dr. Austin Meldon deposed he was called to see the deceased on the morning of the 16th instant. He found her in the spasm of lockjaw. She was actually in the spasm when he entered the room, her body being bent forward. Witness was of opinion the disease commenced late on Sunday night. From examination and inquiry, witness had made up his mind that the disease was caused by a slight wound, inflicted by the needle of a subcutaneous injection syringe. There were numbers of marks over her body, where she had been in the habit of making these injections. The slight wound to which he had referred was made on the previous Friday. He found she had been in the habit of using these injections for years. That morning she told him she had used twelve grains of morphia in one injection, and showed him the papers which had contained the four powders. That was an enormous quantity, a quarter grain being a full dose. The immense quantity she used that morning showed she had been using it for years. She told him that she was in the habit of using, when affected with neuralgia, twenty grains in twenty-four hours. There was no case on record of so much being used.*

Witness asked her why she had adopted that mode of taking morphia. She said, in order to avoid the temptation of taking more of the drug. There was a case of poisoning from morphia in the same way in London, last year, but the quantity was considerably smaller—the dose taken being only one grain four times a day. After

* In my chapter on "The Morphia Habit," I shall show that much larger amounts have been used, and for long periods.

witness saw her, he continued the injections during her spasms, and they relieved her pain, but, of course, the doses he gave were very small, and as the suffering became less, so did the quantity in the injection he administered. He would account for the lockjaw which caused her death by the particular puncture in the skin, as a nerve might have been injured by the entrance of the needle. It was a very hazardous thing for an unprofessional person to use one of those needles. He knew of two cases of lockjaw caused by it. In one of these cases, the patient was very nearly poisoned, for he used when he had no pain a dose which had been ordered him by a medical man when he was in great pain.

Witness made a careful post-mortem examination. He had never seen a lady of that age whose organs were in a more healthy condition. The reason she used it, I may say, was to relieve facial neuralgia, in the first instance, and the habit grew on her. I found the surface of the body punctured in innumerable places with the needle. She seemed as if she had been tattooed.

"*Coroner*. Is there anything else you think it well to tell us? Are you certain she died from traumatic tetanus? ' I am clearly of that opinion, both from the history and condition of the case.'

"*Coroner*. There is one point which I would wish to have cleared up. Several medical men have mentioned to me that it is quite possible she might have obtained, either by mistake or otherwise, strychnia in place of morphia. You are satisfied that is not the case? ' I am perfectly satisfied. I should say that the symptoms of tetanus and strychnine poisoning are the same while the spasms are on.' After the spasms pass away, the patient becomes quite well in strychnine poisoning, but the muscles remain contracted in lockjaw.' "

I have written Dr. Meldon, asking for histories of the two cases of tetanus referred to, but as yet have not had a reply. The *British Medical Journal*, in commenting on this case, says that it has no knowledge of the case referred to by Dr. Meldon as occurring in London, but refers to three cases of death from traumatic tetanus after the hypodermic injection of the sulphate of quinine (*Lancet*, July 6, 1867), and a case of tetanus after the use of morphia, due, probably, to the use of rusty needles (*Lancet*, December, 1876, p. 873).

CHAPTER VI.

THE TREATMENT OF THE OPIUM AND MORPHIA HABITS.

So many of our fellow-men have been, in the past, are now, and will be, bound hand and foot in this terrible bondage, from which they seem utterly incapable of releasing themselves; so many illustrious men have offered themselves as living sacrifices to this false deity; so many lives have been ruined, homes desolated, hopes destroyed, ambitions smothered by it; so many of those dear to us have fallen beneath the shadow of this sickening evil; so rapidly is the habit spreading, that the question of its treatment and cure has become one of momentous importance.

Reading the histories of those cases where a cure was effected some years ago, one's mind is divided between admiration of the great heroism displayed by the unfortunates, and pity for the agony which they were allowed to endure.

Great advances have been made in the treatment of this affection in the last few years, and to-day, although science is unable to substitute for the accursed drug one that can fully take its place, still she can lend a helping hand, whereby the ascent is made easier and more rapid. The suffering incident to the breaking of the habit can, in a great measure, be relieved.

The method strongly advocated and practiced by Levenstein, of Germany, I consider barbarous in the extreme, and dangerous. The following cases, a type of all, selected from those published in his book,* well illustrate this:—

* " Morbid Craving for Morphia." London, 1880.

Mr. von X., sent to the Institution by Professor Westphal, had caught cold while on a journey, which brought on rheumatic pains. To relieve him injections of morphia have been administered since 1872, at first by the medical attendant, and afterward by the patient himself, and in increased quantities, the largest daily dose having amounted to sixteen grains. The symptoms showing themselves in consequence of this use were loss of appetite and sleep, excited condition, emaciation, tremor of the hands.

On October 9, 1875, patient came into the Institution; he had injected morphia for the last time on the morning of the same day. At 10 P.M. patient went to bed and at once fell asleep. At 3.30 A.M. he was sick, felt very weak and prostrate, suffered from twitching in the lower extremities and diarrhœa.

October 10. In the morning patient had five relaxed motions. Frequent vomiting during the whole of the day. Excitement and intense craving for morphia increases hourly. In the afternoon he talks of suicide.

October 11. Patient has had no sleep during the night, but has been frequently sick. Severe vomiting continued until 11 A.M., but stopped entirely during the rest of the day. Patient complains of languid pains in the legs, severe pain in the stomach.

October 12. At 10 P.M. patient suddenly started up, and in a frightened manner asked, several times, "Was not the doctor in the room just now?" only the nurse having been present. Until 12 P.M. he laid in bed quietly without speaking; he afterward raised himself up and screamed out, in great excitement, "Who is that big fellow in the next room? He is so tall that he cannot get through the door! And now he is getting taller still. Now there are several of them; they are ghosts!" His voice was

trembling, his extremities in constant convulsive movements.   He was quieted, but only with great difficulty.
Again and again he raised himself and anxiously looked at
the door.   Temperature, 38.5ᵛ C. (101.3° F.).   In the
morning the patient addressed the superintendent, as he
entered, as follows: " Ah, good morning, dear Emily.   I
am very glad you are coming ! "   While saying so
he was lying down quietly.   Now and then he raised
his head slightly and looked at the wall for a time,
as if observing something, and his lips were moving as if
he was talking to somebody.   In the course of the day he
vomited considerably, several times.   Patient feels very
weak, the speech is unintelligible, the tremor has increased.
He entered into a short conversation, and thinks himself
better than yesterday.   Although very tired he cannot get
to sleep.   Pulse was strong throughout the day.   Auscul-
tation and percussion of the lungs and heart showed a
normal condition.   The bladder was empty and no urine
was passed until 5 P.M.   Several relaxed motions.

October 13.   Toward midnight patient suddenly raised
himself up, looked around, stretched his hands, as if frightened, and called out in a trembling voice, "What do you
want ?   There is the—the ghost ! "   The voice next morning was hoarse, hesitating, unintelligible, devoid of sound.
The features looked worn.   During the whole of the day
there was diarrhœa and vomiting.   An injection of food
into the bowels was given (after Leube).

October 14.   Patient has slept for only a quarter of an
hour in the night.   The other part of the time he was
dozing; vomited four times, and had four relaxed motions.
At 5 A.M. he called out to those watching him during the
night, " Come along, come along, quick, quick ! "   He
gradually lost consciousness, did not move upon being

called. Pulse 40, very small, hardly to be felt; respiration gasping, slow. Hippocratic face. Injected one-fourth of a grain of morphia. Pulse and respiration soon became regular, and he regained his consciousness. There was no vomiting during the day. The voice is still gone, the features worn. Toward the evening the patient had an injection of food (Leube) of sixteen and a half ounces. Great prostration. Skin moist and hot.

October 15. Patient had no sleep, but lay quietly in bed until about 2 A.M., when there was vomiting, oppression, moaning, clonic contractions of the muscles of the face and extremities. Pulse strong.

October 16. He had three relaxed motions, vomiting and bleeding of the nose during the night. During daytime patient felt well.

October 17. Patient has had no sleep during the night, but was quiet; vomiting and diarrhœa. In the day feeling of great weakness. Appetite good.

October 18. No sleep, patient feels thoroughly knocked up. In the afternoon he slept for a short time.

From this time the patient's condition was satisfactory. He slept at first for three hours, afterward for five hours, at last during the whole of the night; the appetite increased considerably, the disposition was changed. He left the Institution on November 21st.

Urine.—The specific gravity of the urine vacillated between 1.019 and 1.029. A precipitation with alkaline solution of sulphate of copper was noticeable only occasionally.

The patient, whom I saw six months after his discharge, has had no relapse.

Mrs. Jane G., a patient of Dr. B. Fränkel, of Berlin, 35 years old, eleven years ago, after having been suffering

with typhoid fever, was afflicted with an abdominal complaint which caused so much pain that the family doctor
had to administer an injection of morphia daily.   Ten
years ago patient married, and has given birth to two
children one five years, the other eight years old now.
The confinements were protracted ; both children died
soon after their birth.   During both pregnancies the
use of morphia was discontinued by the doctor, the same
taking place during several occasional journeys to bathing-
places made by the patient because of her complaint.   For
five years she has injected morphia herself, the largest dose
pro die having been eight grains.

While using the drug, a febris intermittens tertiana
showed itself, two years ago, lasting, with an interruption
of four weeks, until November, 1876.   Regularly at 4.30
P.M. she had shiverings, followed by burning heat, and
ending in perspiration.   The repeated use of quinine, even
a change of air and a sojourn in the country, were unable
to suppress the fever.   Dr. B. Fränkel, who had only for
three months attended Mrs. G., and whom she had never
told of her custom of using morphia regularly, made the
diagnosis of morbid craving for morphia only through considering the intermittent fever.   Apart from the latter, the
use of the drug had brought on the following symptoms :
sleeplessness, headache, principally in the region of the
right occipital nerve, parched mouth, loss of appetite, nausea, sickness, constipation, feeling of oppression, mental
anxiety, palpitations of the heart.   Patient, after having
hardly fallen asleep, wakes up with dyspnœa, which increases to actual fits of choking ; swimming before the eyes
and muscular quivering.   –

Patient is admitted into the Maison de Santé, and the
use of the morphia is stopped forthwith.

October 16. Hardly any sleep during the night; in the morning patient is in a happy temper, makes no complaints. Temperature and pulse normal. In the course of the forenoon there was much perspiration, and patient complained of headache and nausea. Pupils unequal, *the left smaller than the right.* Pressure on the stomach, shivering, yawning; in the afternoon, there is restlessness, stomach-ache, epigastric pain, oppressiveness, much perspiration, shivering. Poultices were applied to the abdomen. Up to the evening she had altogether vomited twelve times and had one relaxed motion. *Pupils unequal, the left wider than the right.* The excitement, in consequence of the pain in the stomach, increases hourly. Patient moves about in bed and moans aloud. Frequent spasmodic yawning. At 9 P.M., a bath with cold douche of a quarter of an hour's duration is given, after which she became quieter for a short time.

October 17. Patient has been sick thirteen times during the night; no sleep; the pains in the abdomen were very severe, having the character of labor pains. Face pale, pulse 64, full, regular. Frequent yawning, burning in the throat, and abdominal pains continue during the whole of the day. Patient looks worn out, is now and then in a half-dozing condition. *Left pupil wider than the right.* Great prostration, great thirst. In the course of the day vomited nine times, two relaxed motions.

October 18. No sleep during the night, restless; complains of tearing pains in the legs, and excessive pain in the stomach; pulse and respiration normal; vomited four times, one relaxed motion, frequent sneezing during the day; she is very sensible to every kind of noise. Frequent retching; severe vomiting ten times. *Left pupil wider than the right.* Patient feels cold. In the afternoon

at 5 o'clock patient lisps, becomes of a pale, death-like
color, is very much oppressed, and loses consciousness;
sinks back on to the pillows with closed eyes.   Pulse 42,
small, irregular.   A quarter of a grain of morphia was at
once given, and repeated after twenty minutes' time.
Right pupil wider than the left.   After a quarter of an hour
patient wakes up, says that she had never felt so well before; takes milk with relish without bringing it up again.
Pulse 60, strong and powerful.   This favorable condition
lasts till 9 P.M., when she again has nausea from time to
time.   At 10 o'clock a bath, with a cold douche, is given.

October 19.   Patient has slept only from 10.30 P.M. to
1.30 A.M.   At this time retching again occurs, vomiting,
prostration, pains in the epigastric region ; hallucinations,
illusions set in, followed by collapse accompanied by the
symptoms already mentioned, and necessitating the immediate injection of half a grain of morphia at 2.20 A.M.,
followed by a weaker injection of a quarter of a grain of
morphia at 4.15 A.M.   Afterwards patient again felt quite
well.   During the forenoon her condition has been good
only at times, the principal complaints being great prostration, impossibility to sleep, pains in the stomach, great
thirst.   *Left pupil wider than the right.*   About 10 A.M.
the sickness increased to such an extent that an injection
of one-fifth of a grain of morphia had to be administered.
Feeling well after it, patient partook of a pint of milk and
soup.   In the afternoon she had some cocoa, which she
has not brought up.   Towards the evening she felt oppressed, which, however, subsided after a warm bath, with
cold douche.   At 9.25 P.M. another injection of a quarter
of a grain of morphia had to be given, on account of symptoms of a collapse showing themselves.   Only three-quarters of an hour afterwards the good effect was visible.

October 20. Patient has slept altogether for five hours during the night, with many interruptions. During the intervals, besides being restless, there was prostration, craving for morphia, nausea, and frequent vomiting, and pains in the stomach. In the morning sneezing and yawning. During the daytime the condition was comparatively good. *Left pupil wider than the right, towards the evening the contrary taking place.* At 8.30 P.M. patient had a bath at 31° R. (87.8° F.) of half an hour's duration, with cold douche.

October 21. Restless during the whole of the night, craving for morphia and increased reflex action. During the daytime the condition of the patient is satisfactory, excepting some yawning, sneezing, and slight prostration.

October 22. No sleep during the night, paroxysms of sneezing and yawning.

October 23. Patient slept for three hours with interruptions. Towards the morning severe sneezing. Appetite good. She had a bath, with cold douche, morning and night.

October 24. Patient was very restless in the night, moved about in the bed, and in the morning was much exasperated on account of the bad night. At 9 P.M. 40 grains of chloral were given in gruel, but were immediately brought up again.

October 26. Patient slept for two hours with interruptions. During the remainder of the night she felt oppressed and had palpitations of the heart. Sneezing the same as yesterday.

October 28. Very restless during the night. Patient had only about one hour's sleep towards the morning, and then felt pains in the lower part of the abdomen.

October 29. Has had hardly any sleep during the

night.   Mental condition nevertheless good ; appetite the same.

October 31.   During the past nights she slept, on the average, for three hours.   Meals are taken regularly. Thirty grains of chloral were given in capsules.   Patient in the daytime complains of labor-like pains in the lower abdominal region.

November 1.   Patient slept for several hours after having taken 30 grains of chloral.   The pains in the hypogastric region are still apparent now and then ; some reddish watery spots show on the linen.

November 2.   The pains have increased in the morning ; the whole abdominal region is sensitive to the touch. Poultices were applied.   *In the middle of the day the menstrual discharge shows itself.*

November 3.   The pains have abated.   Patient has slept for four hours.

November 4.   Menstrual discharge still continues. While it is said formerly to have lasted only for a few hours, it has now lasted for forty-eight hours.   Toward midday the patient left her bed and remained on the couch for several hours.

November 5.   Slept from 11 P.M. till 2.30 A.M.   Patient felt oppressed in the night ; there was difficult breathing, and she could not remain in bed.

Patient leaves the Institution on November 15th, all bodily functions having become regular.   She has up to the present time had no relapse.

In favor of this plan of treatment Levenstein urges the following : " Confidence in the medical adviser is strengthened in consequence of the short duration of the severe symptoms, and the improvement already experienced after a few days ; the patients take courage, look forward to their complete recovery, and submit with patience and resignation to the few days of suffering." And again : "The human organization, as we know from our surgery, midwifery, etc, will, in general, submit more easily to sudden and energetic treatment, even when acting powerfully, than to a milder influence. The gradual deprivation requiring a long time, excites the physical and moral powers to a greater extent, because every dose, smaller than the previous days' quantity, will produce new symptoms of reaction. The constant mental anxiety in which these patients live, while expecting a smaller dose on the following day, makes them fretful and irritable ; their intention of submitting till the end of the cure, and their energy, begin to decline, and they try to evade the treatment. They set up intrigues against the officials and nurses ; they simulate morbid appearances, in order to excite the pity of their relations and friends ; they lose confidence in themselves and in their doctor, whose full and absolute authority is indispensable for the successful treatment of abstinence."

This is certainly true of those cases where the treatment is by very gradual reduction, not at all so of those where it lasts but from four to seven days; ten at the most. The latter plan combines all the advantages of Levenstein's method, and escapes the danger and misery of his, and that of the long protracted course.

In many instances persons addicted to this habit gradually reduce the quantity they have been using to a certain point, beyond which they seem unable to go. Thus, a physician who came under my care reduced his dose from three grains to one-twelfth of a grain in the twenty-four hours, and maintained it at this point for a long time. Finally, however, he returned to the use of the full amount.

He employed the subcutaneous method. The majority of patients express their willingness to be rid of the habit, and do endeavor, up to a certain point, to assist themselves, but at this period will power, naturally weakened, gives way, and good resolves are thrown to the wind. It is at this time that every facility for full control of the patient is necessary, for without it the sufferer will invariably stop treatment, claiming that the suffering is beyond his strength, bemoan his sad fate and return to the old habit with renewed force, exclaiming, with Coleridge's son—

"O woeful impotence of weak resolve."

The treatment of these cases at the homes of the habitués is rarely successful. Some ruse, some strategy, some deception is sure to be practiced, either by the patient, the friends, the relatives or the nurse. Very often the relatives, not understanding the meaning of certain symptoms, distressed beyond measure by the pitiful pleadings of the sufferer, will interpose and at once put an end to treatment, thus unwittingly and with well meaning doing the patient an injury of the gravest kind. For, the treatment, persisted in almost to the point of cure and then abandoned, so thoroughly disheartens the person that it takes a long time for him to make up his mind to try it again.

Those who are the most likely to behave in a manner such as to cause their friends to interfere, from fear of death or insanity resulting, are hysterical females. Absolute committal to a public or private institution, where the nurses can be *absolutely* relied upon, and where obedience to the physician's orders are *rigidly* enforced, is the only rational plan for treating these cases. Then, too, conveniences for baths of different kinds and temperatures, and varied electrical appliances, not found in patient's houses, are necessary; more necessary than drugs.

It is best for the patient, if an adult, or the parents of the patient, if he is a minor, to sign a paper, submitting him fully to the entire control of the physician for ten days. On leaving home his trunk and clothing should be thoroughly searched, and any form of opium or morphine should be removed. On entering the institution in question, the person is allowed from twenty-four to forty-eight hours' rest, to recover from the effects of the journey, to become acquainted with his surroundings, and to allow the physician in charge to judge fully the amount of morphine taken, and the condition of the different organs. The urine should be carefully analyzed and the result saved for future reference. The windows in the rooms are to be firmly secured, and no extra furniture, sharp instruments, or projections from which hanging could be possible, are to be allowed. Low, iron bedsteads are to be preferred, and the room heated by furnace or steam pipes, well protected.

On commencing the treatment, the patient is required to give up all money or valuables he may have about him, for which he is given a receipt. He is then undressed by a nurse, wrapped in a blanket, and taken to the bath-room, where he is given a bath. While this is being done, another nurse, under the supervision of the physician, searches everything the patient has brought with him, including the clothes just removed, and takes away any morphine or opium that may have been secreted, as, also, scissors, knife, needles, etc. The search must be very thorough, as I have known patients to sew packets of morphine into the lining and waistbands of their clothing. The patient is then brought back to his room, the nurse fully instructed, and the treatment by *rapid* reduction commenced.

The physician himself administers the drug, whether by the mouth or by the skin. I usually reduce the quantity used in twenty-four hours one-third, sometimes one-half, the first day. The following case will illustrate the plan pursued :—

Miss B. (sent to me by Drs. Claggett and Walls, of Baltimore), single, aged twenty-seven, born in Virginia. Tall, thin, emaciated. Weight about ninety-eight pounds. Complexion reddish. Dark-brown hair, grayish-blue eyes. Height five feet seven and one-half inches. Extremely nervous temperament. Pupils irregular and contracted. Entered my house for treatment October 16, 1880. Had been taking morphia subcutaneously for eighteen months. It was first given her by a physician, to relieve the intense pain of an attack of pelvic cellulitis. At the end of the treatment, being still a sufferer, she procured a syringe, and continued the injections herself, gradually increasing the dose until the daily amount reached six grains. Sometimes it would be less than this. She was accustomed to take three injections in the twenty-four hours ; one in the early morning, one about midday, and one late in the afternoon. She ceased to menstruate three months after she commenced to use the drug, and had not menstruated up to the time of admission. No history of alcoholic excesses in ancestors. Has one sister who is extremely nervous, and a brother just recovering from paralysis of one side of the face.

Her face is dotted with pustules, as also the chest. Body and limbs marked by cicatrices of old and recent punctures. Appetite fair, bowels constipated. Somewhat lethargic and stupid in mind, but, withal, very nervous. Is despondent, and cries easily, but is very anxious to abandon the habit. There is some vaginismus and spasm of the sphincter of the bladder.

October 15, 1880. Seen by Dr. T. Addis Emmet, who

pronounced her to be free from all uterine and ovarian disease, beyond the remains of an old pelvic cellulitis, by which the uterus was drawn somewhat to one side. He recommended the injection of large quantities of hot water.

October 16. Was given four grains of morphia, subcutaneously, in two doses, one at 7 A.M., the other at 3 P.M. Is feeling very much depressed and homesick.

October 17. Passed a fair night. Very nervous in the morning. Given one hundred grains of the bromide of potassium. Fluid extract of coca (Parke, Davis & Co.), in half-ounce doses, at 9 and 11 A.M., 2, 4 and 9 P.M.; also one-half grain pills of cannabis indica, at 8 and 10 A.M., 3, 5, 7 and 10 P.M. Given two grains of morphia at 7 A.M., and one grain at 2 P.M. More quiet; using beef tea (the juice of steak squeezed into boiling water and seasoned), sherry and bottled beer. Some pain over left ovary. Ice cream and milk for supper. 8 P.M., Pulse 100. Less nervous; more talkative and pleasant; expressed a desire for oysters. Pain over left ovarian region somewhat relieved by poultices. Bowels have not moved for two days.

Urine.—Specific gravity 1.018; reaction acid; color pale amber; odor normal; sediment slight; microscopically nothing; chemically nothing save an excess of phosphates.

October 18. Passed a good night. Pupils of medium size and regular. Craves food. Given two-thirds of a grain of morphia, at 7 A.M. Obliged to draw her water; this had been her habit for a long time. Poultices to left ovarian region. Bromide and other medicines continued as before. Two-thirds of a grain of morphia at 3 P.M.

Rather nervous and restless, but bright.   Appetite good.
Given—

> ℞. Mass. hydrarg.,
>   Ext. colocynth co.,   āā   .   .   gr.iij.
>   At bedtime.

One bottle of beer and some sherry wine.   Two-thirds of
a grain of morphia at bedtime.   Left pupil larger than the
right; both somewhat dilated.   Is suffering slightly from
bromism.

During the day she has taken the following mixture every
three hours, while awake :—

> ℞. Strychniæ sulphat.,   .   .   .   gr.ss
>   Tinct. belladonnæ,
>   Tinct. capsicum,   āā   .   . ℥iij   M.
> SIG.—Take ten drops every three hours, increasing three drops
> each day.

October 19.   Slept well all night.   Two-thirds of a
grain of morphia at 5.30 A.M.   Bright.   Appetite good.
Dressed herself and came down stairs.   Half a grain of
morphia at 3 P.M.   Medicines continued as usual, with the
exception of the bromide, which was given in the form of
an elixir (ten grains to the drachm), one drachm after
each hemp pill.   Pupils regular.   Pulse good.   6 P.M.,
had a severe hysterical convulsion, brought on by talking
and thinking about home.   Given two-thirds of a grain of
morphia.   Some headache.   Bowels have not moved yet.
Given seidlitz powder.

October 20.   Rather nervous during the night.   Tossed
from side to side of bed.   Bowels moved naturally about
2 A.M.   Two-thirds of a grain of morphia at 6 A.M.   Pupils
contracted and even.   Slight headache.   Restless.   One-
third of a grain of morphia at 3 and 9 P.M.   Much headache.

October 21.   Bowels moved naturally during the night,

One-third of a grain of morphia at 6 A.M; one-third of a grain at 2.30 P.M., and one-sixth of a grain at 6 P.M. Appetite excellent. Took some exercise in the yard. Hysterical attack at 5 P.M., after reading a letter from home. Given twelve grains asafœtida. Bromide eruption on chest and back. Urine slightly albuminous and containing a trace of sugar.

The spasm consisted of moaning, with greatly labored breathing, the moan being made at each full expiration. There was trembling of the hands and twitching of the muscles, accompanied by a feeling of suffocation. Pulse small and frequent. At 6.30 P.M. was sufficiently recovered to eat a hearty supper. During the evening she felt well. One-sixth of a grain of morphia at 9 P.M. Ordered half an ounce of tincture of hyoscyamus at bedtime.

October 22. Restless all night. Hyoscyamus did not produce more than two hours' sleep. One-sixth of a grain of morphia at 6 A.M. Very nervous and restless all morning. One-sixth of a grain of morphia at 3 P.M. Complains of pain in body, limbs and side. Given a very hot sitz-bath, followed by belladonna plaster to ovarian region. One-twelfth of a grain of morphia at 6 P.M. Less nervous at 7 P.M. Vomited once. One-sixth of a grain of morphia at 9 P.M. Still restless. All medicines continued. Pupils dilated and irregular ; right larger than the left.

October 23. Passed a sleepless night. Very nervous in the morning. One-sixth of a grain of morphia at 6 A.M. Distressing nausea and vomiting. Given ice and wine. In the afternoon there were some symptoms of collapse. One-twelfth of a grain of morphia at 1 P.M., one-half grain at 4 P.M., one-twelfth grain at 5 P.M., and one-sixth grain at 9.30 P.M. Towards night the nausea ceased. Complains of severe pains in limbs and back. Is thoroughly rubbed with whiskey and water. Pupils contracted. 4 P.M. slight symptoms of collapse. Gave stimulants. In place of belladonna, strychnia and capsicum mixture, ordered the following :—

℞.	Strychniæ sulph.,	. . . .	gr.ss.
	Tinct. belladonnæ,	. . . .	ʒ ij
	Tinct. lobeliæ,	. . . .	ʒ j
	Tinct. stramonii,	. . . .	ʒ j
	Tinct. capsicum,	. . . .	ʒ ij. M.

Sig.—Take twenty drops every three hours, increasing three drops daily.

Continued the other medicines in the same way as before. Urine passing freely. Albumen one-twentieth of the bulk. No sugar. Hysterical attack at 9 P.M.

October 24. Slept well all night. Given four ounces of milk and lime water every two hours; also some milk punch. Severe hysterical tetanoid seizure on rising from commode. Lasted over an hour. Great difficulty in overcoming spasm of respiratory muscles. Expiration loud and forcible. Hands clenched. Fingers and toes forcibly flexed. Clonic spasm of muscles of face and jaw. Face purple, veins distended, pulse at wrist hardly perceptible. She was put fully under the influence of ether, with good result. Complained of headache, nausea and soreness of muscles, after the effect of the ether wore off. Says that when she stood, a pain darted into the left ovary, and threw her into the spasm. Medicines continued. Gave hot hip bath, followed by cold douche, and the interrupted current along spine and over ovary. Also massage, which greatly relieved pains in limbs and soreness of muscles.

12 M. Feeling bright and cheerful. Took some bottled beer. Bowels have not moved for two days. Poultice over abdomen. One-twelfth of a grain of morphia at 3

P.M., also one-twelfth grain at 8 P.M. This was the last dose of morphia. At 9 P.M. another hysterical spasm. Again etherized. Soon after this the bowels moved.

12 midnight. *Left* pupil larger than the right. Slept from 2 to 5 A.M.

October 25. Very nervous and restless. Complains of constriction of chest by an imaginary band. Dyspnœa. At 6 A.M. another spasm on rising from commode. Again resorted to ether, after three hundred and thirty grains of bromide of potassium in an hour and a half's time had failed to relieve. A few whiffs of ether were sufficient. Some nausea from the ether. 9 A.M., symptoms of collapse that yielded readily to brandy. In a quiet state all day. Stimulants and liquid nourishment every hour. Some vomiting, sneezing and hiccough. Diarrhœa during the day. Face purplish red. Respiration 36; pulse 120. Stupid but not sleeping.

6 P.M. Improving. Recognizes faces; talks rationally, but speech thick. Takes food well and seems to enjoy it. *Double Vision.* 9 P.M., restless. Wants to get up and walk about the room. Given a subcutaneous injection of water. Slept for two hours. Medicines continued.

October 26. Yawning, gaping and sneezing. Complains of pain in limbs, also of vesical and rectal tenesmus. An examination reveals a small hemorrhoid. Given a suppository containing half a grain of extract of belladonna, which seemed to afford some relief. Pulse stronger. Dyspnœa less marked. Given a hypodermic injection of water at 8 A.M., 3 and 9 P.M., much to her relief. Doubts as to whether there was morphine in the solution used, were dissipated by letting her touch her tongue to the needle, which had been previously dipped in tincture of nux vomica.   Complexion clearer ; eyes bright ; speech not so thick.   Strychnia mixture stopped.

8. P.M.   Great rectal tenesmus, with small, offensive stool.   Given one grain of extract of belladonna, by suppository, and a capsule containing—

R. Mass hydrargyri,
Ext. colocynth co.,   āā   .   .   . gr.iv

The mouth tasting badly, she was given the following as a wash :—

R. Sodæ hyposulphite,   .   ℥ss
Tinct. kino,
Tinct. myrrhæ,   āā   .   .   . ℥j
Ol. gaultheriæ,   gtt.xvj
Aq. rosæ, ... ad   ℥ iij.   M.

SIG.—Teaspoonful in half a glass of water.

Pains in limbs and body still severe.   Much relieved by a hot bath, followed by the cold douche.   Hemp pills stopped.   Stimulants and liquid food continued.

State of partial collapse at 5 P.M.   Relieved by the free use of stimulants.

October 28.   Slept well.   From this time she improved rapidly.   She was put upon cod-liver oil, cream, milk, generous diet and the following tonic :—

R. Strychniæ sulph., ... gr.$\frac{1}{24}$
Tinct. cinchona co.,   .   .   . ℥ij
Three times daily.

In nine weeks' time her weight increased from about 100 to 146 pounds.

The bromide eruption, which was severe, was treated with a wash of sodæ salicylate, ten grains to the ounce.

November 11.   Some puffiness of abdomen and pain in left ovarian region.   Given aloes, iron and myrrh, until bowels were moved freely.

November 17.   *Menstrual flow established.*   Passed

through period with but little pain. Some leucorrhœa followed. Partly relieved by injections of infusion of white oak bark.

December 11. Again unwell. Less pain.

December 23. Discharged well. Absolutely no craving for morphia.

In this case the hysterical element was decided, and somewhat modified the treatment.

The symptoms that follow the sudden deprivation of morphia are so vividly and well described by Levenstein, who has treated many cases in this way, that I cannot refrain from giving his own words :—

"Although persons who suffer from morbid craving for morphia show different symptoms, some of them beginning to feel the effects of the poison after using it for several months, while others enjoy comparatively good health for years together, there is no difference between them as regards the consequences upon the partial or entire withdrawal of the narcotic drug.

"In this respect they are all equal. None of them have the power of satisfying their passions unpunished.

"Only a few hours have passed since using the last injection of morphia, and already the feeling of comfort brought on by the action of the drug is passing off. They are overcome by a feeling of uneasiness and restlessness ; the feeling of self-consciousness and self-possession is gone, and is replaced by extreme despondency ; a slight cough gradually brings on dyspnœa, which is increased by want of sleep and by hallucinations.

"The vaso-motoric system shows its weakness by abundant perspiration, by the dark color of the face, which replaces the pale condition apparent during the first few days.

"Flow of blood to the head and palpitation of the heart, with a hard pulse, soon show themselves. The latter symptom often disappears suddenly, and is replaced by a slow, irregular, thread-like pulse, which is the sign of the beginning of a severe collapse.

"The reflex irritability increases, the patients begin to sneeze and to have paroxysms of yawning; they start if any one approaches them; touching their skin causes crampy movements or convulsions; the trembling of the hands, if not already evident, now becomes distinctly perceptible. The power of speech is disordered; lisping and stammering take place. Diplopia, and disorders of the power of accommodation, frequently accompanied by increased secretion of the lachrymal glands, show themselves. The patients are overcome by a feeling of weakness and total want of energy, and are thus compelled to lay in bed.

"Neuralgic affections of various parts of the body, pain in the front and back of the head, cardialgia, abnormal sensations in the legs, associated with salivation, coryza, nausea, vomiting and diarrhœa, tend to bring them into a desperate condition.

"Some persons will bear up with fortitute under all these trials; they will quietly remain in bed, and will endure the unavoidable suffering, hardly uttering a complaint. Of the others, although the great minority of them sleep and doze during this trying time, some can find rest nowhere; they jump out of bed, run about the room in a state of fear, crying and shrieking. Gradually they become calmer, although occasionally their excitement increases. A state of frenzy, brought on by hallucinations and illusions of all the sensitive organs, at last causes a morbid condition, to which I have given the name of delirium tremens, resulting from morbid craving for morphia, it being similar to that caused by alcohol. Some of the patients, however, will be found walking about in deep despair, hoping to find an opportunity of freeing themselves forever from their wretched condition.''

The patient whom I treated by the sudden deprivation of the drug was a married lady, aged forty-five, stout and flabby. She had been using morphia by the mouth for nearly sixteen years, in the last year taking from seven to ten grains daily. I saw her one afternoon and commenced treatment the next morning. The drug was taken away at once. That day she began to complain of salivation, dyspnœa, intense grinding pains in the calves of the legs, headache, pain about the heart and a strong desire for morphia. The first night she did not sleep at all. Was feverish and restless, tossing from side to side of the bed, moaning, occasionally crying out. Toward morning she became delirious, ran about the room, screamed, attacked the nurse, attempted to jump out of the window and battered at the door. By 8 A.M. she had sunk into a stupor, from which she was occasionally roused by fits of sneezing. A hard, dry, hacking cough supplemented the dyspnœa. Nausea was intense and vomiting frequent. She, on the third day, knew none of those about her, saw imaginary men and animals, wept, laughed, moaned and muttered incoherently. Sordes formed upon teeth and lips; the fæces and urine were passed in the bed. In this state she remained for six weeks, in spite of every effort to arouse her. All day in a typhoid condition; at night staggering about the room, screaming, crying and attacking the nurse. In the seventh week she began to recognize faces, although her hallucinations and delusions continued. She very slowly regained strength. Pains persisted in the limbs.

There was trembling of the hands and tongue, inability to read, and she would cry at the most trivial thing. Her nights were filled with terrible sights and dreams, the memory of which lasted her the whole of the following day. Several times there was severe collapse, necessitating the free use of stimulants and an occasional small dose of opium. During all this time tonics, concentrated food, baths and electricity were given. *It was fully four months before her mind regained its balance.* She was seen once, in consultation, by Dr. G. H. Wynkoop, of this city.

That was my first case, and I shall never try the experiment again. I know of several instances where physicians have tried this plan and abandoned it.

Her urine contained both sugar and albumen in large amount. This afterward wholly disappeared.

Here is further testimony as to the efficacy of the treatment by sudden deprivation :—   •

"Dr. Osgood (*Quart. Jour. Inebriety*, June, 1879,) has, in a hospital, during the past two years, treated 800 cases of opium inebriety. His plan, in general, is: (1) The absolute and total discontinuance of the use of opium from the beginning of treatment. (2) A trusty attendant to be with the patient day and night for the first three days. (3) Chloral hydrate for the first three nights, if required. (4) Good food, milk, raw eggs, brandy (in some cases), chicken broth. (The above to be taken in small quantities.) (5) In diarrhœa give two-drachm doses of a mixture of equal parts of tincture of catechu and tincture of ginger. (6) Vomiting will frequently yield to bismuth in fifteen-grain doses; and in some cases a single dose of calomel has acted like a charm. Ice is of advantage in some cases. (7) Throughout the entire treatment it should be remembered that the patient is below par, and requires tonics. Quinine and tincture of iron have a prominent place in our list. (8) The patient should expect to suffer more or less for the first three days, and should make himself a prisoner for that time. By the fourth day there is usually marked improvement. (9) Usually by the sixth day all desire for opium is gone. The patient then requires a change of air and surroundings, and tonics for a few weeks. Out of one hundred cases thus treated there was but one death and that from apoplexy."

However well this plan may answer for the Germans and Chinese, it certainly is a dangerous and barbarous practice when applied to American and English people.

The dangerous collapse that so often occurs in the carrying out of this method Levenstein combats by small, subcutaneous injections of morphine, stimulants, dropping ether on the skull, etc.

CHAPTER VII.

AGENTS EMPLOYED IN THE TREATMENT OF THE OPIUM AND MORPHINE HABITS.

Limited space forbids my illustrating each modification of treatment with pertinent cases, and I content myself with discussing the various drugs and agents used, and the indications calling for their employment. In this way, too, the reader will become acquainted with the various symptoms that arise in the course of rapid deprivation.

BATHS.

Baths of various kinds are the mainstay in the treatment of most cases of this kind. They are used for four purposes: (*a*) To allay excitement. (*b*) To relieve pain. (*c*) To equalize the circulation, and (*d*) To procure sleep. They succeed when drugs fail.

On the third or fourth day of treatment it is usual to have, especially in very nervous persons, considerable excitement, reflex nervous irritability, not so great, however, as when the drug is suddenly taken away. The violent delirium tremens spoken of by Levenstein I have never seen, save in one case, which I treated according to his method. When delirium occurs at all, it is usually of a low, muttering, harmless kind. For its treatment and that of the severe nervous irritability and restlessness sometimes seen, the hot bath, followed by the cold douche, is an excellent remedy. If the patient is strong I let him remain in it for twenty minutes; if weak, from five to fifteen minutes. A good reaction is had and the quieting powers of the bath enhanced by pouring one or two pails of cold water over the patient's head and shoulders, or directing upon him for a moment a number of fine jets of cold water, with considerable force. The nozzle should be so arranged that from twenty to forty jets may play at the same time. After this the body should be thoroughly rubbed with a hair glove, until the surface is in a glow, and the patient then be put to bed, where gentle perspiration usually follows, with considerable relief to the pain in the limbs, shivering and sneezing ; oftentimes a quiet sleep, lasting from thirty minutes to two hours, follows.

A rapid hot bath, followed by the douche or spray, and that by thorough rubbing, I find an excellent measure in the after-treatment of these cases, it hastening the breaking down of old and stimulating the growth of new tissues, as also the different secretions.

The temperature of the bath should never be below 112° Fahr.

In some neuralgic patients, the cold douche sometimes causes pain afterward. In these cases it should be omitted, the hair glove being used freely in its stead.

Several baths may be given in the course of a day, but I rarely find it necessary to give more than two or three in the worst cases.

The bath, too long continued or too often given, will sometimes greatly weaken the patient. In debilitated subjects and those suffering from pelvic pains I find the hot sitz bath, followed by thorough rubbing and kneading of the whole body, often gives great relief.

For persistent insomnia I find the cold pack a most serviceable agent. The patient is wholly undressed, a sheet is wrung out of cold water, and he is wholly enveloped in it up to the chin. The sheet should be folded under the feet and tucked in evenly and closely about the neck. He is then closely wrapped in from four to six blankets, and in the majority of instances will soon fall asleep, a gentle perspiration breaking out all over him. He may be allowed to remain in this an hour or even two hours.

Excitement and nervousness are calmed by it, the restlessness and pain disappear, and the sufferer often remarks how much good it is doing him.

In removing it, the whole matter should be done rapidly, under cover of a heavy blanket, and the body be thoroughly rubbed with a sponge dipped in whiskey and water.

ELECTRICITY.

This is also a valuable agent in the treatment of the pains, that are usually distressing. Sometimes the interrupted, sometimes the continuous current acts the best. As a rule, however, a powerful continued current (from ten to forty cells) is the most serviceable. The electrodes should be kept continuously moving up and down the limbs, and the current be occasionally reversed.

Dr. William F. Hutchinson relates the following interesting case* of cure by electricity: "About the first of January of the present year, I was requested by Dr. O. C. Wiggin, of this city, to see with him a lady supposed to be suffering with cerebral congestion in an advanced stage.

" A visit to Mrs. S. revealed the following history: Age thirty-nine; married; one child, aged six; and has had one miscarriage; weight about 150 pounds; and general appearance of contour and skin good. Patient kept up a low moaning, answering most of my questions intelligently, then relapsing into a semi-unconscious condition. Pulse 100, compressible; temperature 99°; no loss of

* (N. Y. *Med. Record*), *Southern Med. Record*, Sept. 20th, 1879.

control of evacuations; conjunctivæ congested and pupils contracted closely; perspiration starts upon the smallest exertion, which also causes pain in abdomen and excites vomiting, which has lately become persistent, accompanied with intense thirst. Hands and feet cold, with shriveled palms and plantar surfaces. No difference in temperature of head and axilla.

"Ophthalmoscopic examination gave retinal and choroidal congestion, with venous enlargement, slight optic neuritis and choked disk.

"There was constant pain, and sense of fullness in frontal region.

" The only family history that could be obtained bearing upon the case was the death of one sister, a year ago, from acute brain inflammation, the remainder of the immediate family being still living and in good health; and the present condition appeared to be the culmination of six years of almost constant pain and general nerve exhaustion, following the birth of a child, and aggravated by a subsequent miscarriage.

" At this visit no suspicion was entertained by me of any opium habit, and the case was diagnosed as passive cerebral congestion, dependent upon general neurasthenia.

" The next day Dr. Wiggin called and gave me the following additional items, which at once placed the case in its proper light and gave the key to many of the symptoms before cited. After her confinement, which was a long and painful one, she suffered severely from wandering pains in back and hips, for which her attending physician at the time ordered tincture of opium applied externally, giving at the same time ten drops by the mouth, and the ground was broken for the building of the habit. The dose steadily increased until she came under the charge of Dr.

Wiggin, some six months previous to my seeing her, when she was taking four ounces of laudanum daily, internally, besides continuing external applications as before. Attempts were made to stop the pernicious habit, but it was too late for wise counsel to avail, and the usual cunning of opium eaters procured for her the drug in spite of every effort of both husband and physician.

"All forms of concurrent medication had been faithfully tried, but nothing was of use except the opium, to which it became absolutely necessary to resort occasionally, as without it the poor lady would arouse the neighborhood with agonizing screams and cries.

" At this juncture, as a forlorn hope, it was decided to essay galvanism, hoping that its great vitalizing power might aid in restoring tone to the exhausted nerve centres. At my suggestion, Collis Brown's chlorodyne was given in place of laudanum, and produced the same effect with an ounce per diem that the four ounces of the former had done.

"Central galvanism was applied, with a twenty-four cell Bartlett battery, using six cells from the cilio-spinal centre to the forehead, with a downward current; then from the cervical vertebræ to the solar plexus, with an ascending current, each lasting six minutes, or until the skin was thoroughly reddened under the negative carbon point. For the first few days applications were made morning and night. In a week the vomiting had ceased and consciousness returned, and the evening sitting was omitted. After a month the dose of anodyne was gradually decreased, but with every diminution the nausea returned, and nothing but a return to the old dose would avail. But her condition was very much improved. She slept better, the eyes were normal as to color and the palms were no longer dry.

At the close of the second month she was able to sit up, and the dose of anodyne was steadily cut down without the patient's knowledge, by adding to the chlorodyne a sufficient quantity of flavored treacle to replace each dose taken, until at that time an ounce would last three days.

" Her general condition was greatly improved, and she began to take interest in her surroundings. In two months more she commenced to go out, and came to my office for treatment, when I changed the current to the Siemens and Halske cabinet cell, which, with its low tension and perfect capability of control, I regard as the ideal battery for central galvanism. There was no further trouble, and to-day, June 21st, the lady is quite well, attending to all her household duties, not having tasted opium in any form for seven weeks, and expressing unbounded delight at being free from the terrible habit which had so long been her master.

" The *rationale* of the action of galvanism in this case is difficult to understand. When the circuit was closed over the superior sympathetic ganglion, Dr. Wiggin and myself distinctly observed a sudden wave-like contraction of the distended retinal veins, which resumed their size in a few moments after the stimulus was removed. But, after some weeks' treatment, these veins became normal, and the intraocular congestion had disappeared *pari passu* with the cerebral symptom, and having repeatedly witnessed the same phenomenon in other cases, I am led to believe that the galvanic current has a direct tonic influence upon the vaso-motor system, which accounts for the occasional surprising results obtained in cases of cerebral congestion. With the advent of increased nerve circulation came an absolute horror for the drug, and it is not easy to know to what to attribute the increase of strength of will up to the point of totally dispensing with it of her own accord, unless it be to some change in mental power, due to increased nerve tone, the direct result of what I have before termed the vitalizing power of the galvanic current. Faradism was not at any time employed.

"Dr. Wiggin gives full credit to the special treatment for the cure of the case."

DRUGS.

The *bromides* I have found to be very valuable, especially where there is a hysterical tendency. The bromide of sodium is said to disorder the stomach less than any of the others. To be of any service they must be given in large doses. Dr. J. B. Mattison gives sixty grains, three times a day. I have given as high as three hundred and thirty grains in one hour and a half, and often in one hundred grain doses, and have never yet seen the least harm result, but very decided good. Wyeth's elixir of the bromide of potassium is an elegant preparation. Its strength may be increased at will. Where there is too decided irritation of the stomach I give the drug by the rectum, in milk. Occasionally a large dose of the bromide will quiet persistent nausea. It certainly modifies the excitement and reflex irritability, as also the headache and salivation.

Coca I have tried in several cases, but cannot agree with Prof. Palmer as to its peculiar adaptability to these cases. As a nerve tonic it is a valuable adjunct to other treatment, but beyond that nothing.

Strychnia, capsicum and *belladonna* I use in large doses, or small doses often repeated. They are all heart stimulants, and the two former decided nerve tonics, aside from their action on the stomach. I vary the quantity of each ingredient to suit each case. The use of capsicum was first suggested by its successful employment in delirium tremens.   A nervous tonic may at the same time be a sedative to nervous irritability.   I gradually increase the dose as the symptoms are wont to increase, during three or four days, and the drugs thus given seem to have a better effect.

If the nervous irritability is extreme, especially if there are hysterical symptoms or twitching of muscles, or spasm of the rectal or vesical sphincter, I add *lobelia* and *stramonium,* which, besides relieving these symptoms, are excellent in allaying the nausea and vomiting.

Cannabis indica is an excellent agent to cause tranquillity and destroy hallucinations.   It seems to take the place of the morphine to a certain extent.   It, too, must be given in large doses.

Hyoscyamus and *hyoscyamia* I have tried, and never derived much advantage from them.   The former usually increases or produces diarrhœa.   My failure to obtain good results has not been due to using too small doses.

Chloral, except in small doses, is a dangerous remedy, and one with which, if a good effect is had, the patient is apt to fall in love.   It is not of much service as a sleep producer, save in dangerously large doses.   Levenstein found that in some cases it produced violent excitement. I have seen a ten-grain dose produce urticaria that tormented the patient for several days.

Hydrocyanic acid was first recommended in the treatment of this affection by W. C. Blalock.*   He claims that it supplies the place of morphine or opium.   It does not do so.   His formula reads as follows —

    ℞.   Acidi hydrocyanici dil.,   .   .   .   gtt.xlviij
        Syr. simplicis, ...   ℥ ij
        Aquæ, ...   ℥ j.   M.
Sig.—A teaspoonful at 7 A.M., 12 and 8 P.M.

* (Atlanta *Medical Journal*) *Physicians' Monitor,* 1878.

As a quieter of gastric irritation and nausea, and as a nerve stimulant, it is an excellent adjuvant in certain cases.

Lupulin and *Lactucarium.* I have not had sufficient opportunity to test these drugs fully, but in one case, which I have under treatment now, they seem to take the place of morphine with excellent results, the patient going twenty-four hours with but a small amount of discomfort on a minimum amount of morphine and large amounts of these drugs. In three days I have been enabled, by their use, to reduce the amount of morphine from twelve grains to half a grain, in the twenty-four hours, the patient being up and about. Whether the effect will hold, and whether these drugs can be easily abandoned, remains to be seen. It is in drugs of this class that we must look for a substitute for opium and its alkaloid. With other treatment we merely combat the symptoms that arise.

The fluid extract of lupulin may be used in from one to four drachm doses, by the mouth, or from thirty to sixty drops hypodermically. Lactucarium is best given in powder, in half drachm doses.

I have also used the extract of lettuce (English) in ten-grain doses, every two hours.

DIET AND STIMULANTS.

In the management of these cases diet is as important as medication. With the plan which I am now pursuing I rarely have nausea lasting longer than a few hours, a very important point, both as regards the comfort of the patient and the possibility of full nutrition. Diarrhœa, too, so common in Levenstein's plan, rarely proves troublesome, is often entirely absent, and I am occasionally obliged to use laxative medicine. If there is any tendency to relaxation of the bowels I interdict the use of fruit, of which these patients are very fond, and give subnitrate of bismuth, in large doses.

Beefsteak, beef broth, chicken soup, mutton broth, wine jelly, eggs and plenty of milk are called for. With the milk a little lime water should be used. If the digestion is slow and labored, from ten to twenty grains of pepsine should be given after each meal. If there seems to be decided gastric catarrh nothing acts so well as pills composed of—

> ℞. Argent, nitrat.,   .   .   gr.$\frac{1}{6}$
> Creasote, ... gr.ss.
> One, one hour before meals.

If there are symptoms of collapse, although this is unusual, wine should be given freely, or whiskey hypodermically, three or four syringefuls at a time. Milk punch is an excellent stimulant and food, and egg-nog good, when well borne by the stomach, which is not always the case.

The best wine for these patients is sherry, and it should be better than that found in drug stores and bar rooms. Port is usually largely adulterated. Claret causes acidity. As a rapid stimulant the best champagne is excellent. There comes a little apparatus that can be screwed through the cork of the bottle, and by turning a faucet as much or as little can be drawn as is wanted. It saves wine and keeps it from staling.

Bottled beer is of service in some cases. The purest only should be used. It may be kept from staling in the same manner as the champagne.

Alcoholic stimulants should be continued only so long as is absolutely necessary, which is never more than ten days. This is very important, for it must be borne in mind that these patients are prone to go to excess with stimulants, as well as narcotics.

AFTER TREATMENT.

The thorough cure of an opium or morphine habitué does not consist alone in stopping the use of the drug. This is certainly a very important step, but there yet remains much to be done, especially in those cases where the person became habituated to the use of the drug through its employment in some extremely painful neuralgic affection. This disease may still persist or return in full force on the withdrawal of the drug.

The use of the drug being discontinued, attention should be at once directed to the treatment of the original affection *or of any disease that may lead the patient to again return to the use of the drug.*

There are, too, certain troublesome symptoms, about some of which I have already spoken, that come on a few days after the opium or morphine is stopped. The most troublesome of these is "soreness of the throat," consisting in congestion of the pharynx, tonsils, larynx and vocal cords. There is a sensation of tickling and a short, dry, annoying cough, that comes on in paroxysms and is especially troublesome at night. It is best treated by the local application of nitrate of silver (twenty to forty grs. to the ounce) and five-drop doses of benzoic acid given several times daily.

The insomnia is usually relieved by exercise better than by drugs, although I have often found bromide of potassium and hyoscyamus of service. On no account use chloral hydrate, or give stimulants at bedtime. Exercise is by far the best remedy. Walking, running, making garden, etc., sufficient to tire the patient out thoroughly, is an almost certain sleep producer, appetizer and hastener of tissue metamorphosis. The cold pack or a hot bath at bedtime often proves of service.

The appetite usually needs no stimulation by drugs. It is, as a rule, ravenous for the first few weeks after the morphine is stopped, so much so that ten-grain doses of pepsine will often be found necessary after each meal, the patients usually over eating. As a rule, I give the following mixture, as much for its effect on the nervous system as for the appetite : —

     ℞. Strychniæ sulph.,   .   . . gr.ss *vel* gr.j
         Tinct. gentian co.,
         Tinct. cinchona co.,   āā . .   ℥ ij.   M.
   Sig.—One drachm half an hour before meals.

If anæmic, nothing acts so well as the muriated tincture of iron, which usually contains a trifle of arsenic, in from fifteen to forty-drop doses, three times a day, *after meals*. It should be given in half a goblet of water. Cod-liver oil is also of great benefit, especially if there is combined with it minute doses of the iodide of arsenic.

Seminal emissions and priapism in males will right themselves.

Females who suffered from leucorrhœa before they commenced using opium or morphine, and who stopped during the time they were taking it, as, also, those who have never had such a discharge, are usually troubled with a white, ropy matter passing away from them in large amount. This is sometimes accompanied by severe, labor-like pains. Vaginal injections of a strong infusion of white oak bark are called for, to be followed by the use of the following on the top of pads made of borated or absorbent cotton : —

     ℞. Tinct. iodin comp.,   .   .   . ℥ ss
         Glycerin, ... ℥ ij.   M.

These pads are made a little larger than a lemon, with a pedicle twisted with fine cord. They are medicated on the top, the sides smeared with vaseline, and are then pushed up against the mouth of the womb.   A small piece of cord is left hanging, so that they can be easily removed. They are to be introduced in the morning, and removed at night.

Neuralgic pains in the limbs and body are best removed by the hot bath and massage.

The menses may come on irregularly, that is to say, not at the period at which they usually came before addiction to the habit caused their suppression.   Pain in the pelvic region, fullness of the abdomen, and a general sensation of puffiness, indicate their coming.   On the appearance of these symptoms I give the following until the bowels are freely moved, at about which time the menses usually appear:—

> ℞. Ferri sulphatis,   .   .   gr.ij
>   Pulv. alöes soc.,
>   Pulv. canella,   āā   . gr.iv.   M.
> In capsule every night or morning, or the pills of Oppolzer.*

The period is finally established at its proper time by coming on a few days sooner each month until the normal date is reached, after which there is rarely any change. The second and third menstruation is much more abundant and less painful than the first.

Dyspepsia and "biliousness," which are sometimes present, are best overcome by pepsine, and the following, every third night, to be followed in the morning by some laxative mineral water:—

> ℞. Mass hydrargyri,
>   Ext. colocynth, co., āā .   .   . gr.iij.   ˙M.
> It may be made in pill form or be put in a capsule.

---

* H. C. Wood; "Materia Medica and Therapeutics," p. 440. Phila., 1877.
> ℞. Ferri sulphatis,   .   .   gr.xx
>   Ext. alöes, aq.,   .   ℨj
>   Ext. taraxaci, ...   q.s.   M.
>   Ft. pil. No. 60.
> Sɪɢ.—Two, night and morning.

Finally, *plenty of exercise and fresh air*, baths, regular habits of life, and treatment directed to those complaints, real or imaginary, if still existing, for the relief of which the patient first took the drug.

The rapid way in which these patients gain flesh is sometimes astonishing. A gain of fifty pounds in a month's time is not unusual.

PROGNOSIS.

The prognosis is, in the majority of cases, good. A cure can be effected in any case, provided the directions followed are faithfully carried out, especially in a private institution, where absolute control of patient and nurse is possible, and where the number of patients is limited.

Levenstein says : —

" To treat morbid craving for morphia with success, it is necessary to decide the principal question, namely, whether each individual patient does or does not suffer from pathological complaints or chronic disorders requiring narcotics for their relief. If he does, it is only necessary for the doctor to deprive the patient of the morphia syringe and to inject personally, if his time permits of so doing, a dose which he thinks sufficient, or else to give the narcotic internally.

" Furthermore, we exclude from the treatment all patients weak or exhausted through bodily or mental affliction. It is all the same whether the prostration is caused by night duty, distress, illness, childbed, want of proper food, flooding, etc.; only those conditions of weakness following upon the poisoning with morphia constitute no counter-indication, as they disappear in consequence of the deprivation. Only such individuals, therefore, are suited for the treatment who have continued the injections while in perfect

health, the former morbid appearances for which they at
first administered them having disappeared.''

From these statements I must heartily dissent.   No
organic disease, no affection of the general system, save
those that are sure to prove fatal, as cancer and the like,
justify continuance in this habit.   In almost every instance
the accompanying ill effects will be found to be due more
to the abuse of morphine than to the disease itself.   Indeed, its continuance may place the patient in a condition
that will preclude the possibility of recovery.

Patients with one foot apparently in the grave, when
deprived of their morphine, take a decided turn for the
better, and regain fair, if not perfect health.   A fatal
ending sometimes follows the waiting and attempting to
build up the health before commencing the treatment of
the habit.

As to the advisability of treating debilitated and exsan-
guinated patients by the method of sudden deprivation, I
fully agree with him, for the trial is as much as a robust
and healthy individual can bear.   By substitution and
gradual deprivation, however, it can be accomplished
safely and satisfactorily.

The danger of a return to this habit decreases with each
year, each month, each day of abstinence.   It is greater
in neurasthenic patients and those who have taken large
amounts, or who have not received proper after treatment.

Under no circumstances should these people be given
opiates in any form or for any complaint, *save when life
can be saved in no other way*, for a relapse is almost certain
to occur if this is done.   Witness the case of the literary
gentleman already referred to.   *A single dose may undo
the work of years.*

In conclusion, I feel that I am warranted in again insisting upon the growing evil of indiscriminate and careless use of the hypodermic syringe. The following, from my work upon the hypodermic use of morphia, well expresses my feelings in this matter.

"Some of my correspondents, men of ability and in large practice, express themselves as very skeptical of the truth of the statement that the morphia habit has ever been formed by the use of the drug hypodermically. Testimony from all parts of the civilized world settles this matter beyond question. Bartholow, from whose excellent little work I have so often quoted, says :*

"'The introduction of the hypodermic syringe has placed in the hands of man a means of intoxication more seductive than any which has heretofore contributed to his craving for narcotic stimulation. So common now are the instances of its habitual use, and so enslaving is the habit when indulged in by this mode, that a lover of his kind must regard the future of society with no little apprehension. It may well be questioned whether the world has been the gainer or the loser by the discovery of subcutaneous medication. For every remote village has its slave, and not unfrequently several, to the hypodermic syringe, and in the larger cities, men in business and in the professions, women condemned to a life of constant invalidism, and ladies immersed in the gayeties of social life, are alike bound to a habit which they loathe, but whose bonds they are powerless to break. Lamentable examples are daily encountered, of men and women, regardful only of the morphia intoxication, and indifferent to all the duties and obligations of life, reduced to a state of mental and moral weakness most pitiful to behold.

"'Usually the habit is formed in consequence of the

* "The Hypodermic Method," etc., p. 90.

legitimate use of the hypodermic syringe in the treatment of disease. Employed in chronic painful maladies for a long period, it is discovered, when an attempt is made to discontinue the injections, that the patient cannot or will not bear the disagreeable, even painful, sensations which now occur. More frequently, when the injections are to be used for a long time, the patient is unwisely intrusted with the instrument, and taught all the mysteries of the solutions and the mode of administration.'

"A study of the opinions and cases that I have collected and presented in the preceding chapters furnishes abundant food for thought; the more so, as nearly every physician in the world is using a hypodermic syringe with more or less frequency. Many are unquestionably using this instrument too often; are using it in cases where the same, or other remedies, by the mouth, would be equally efficacious, and certainly safer. Far be it from me to condemn the use of an instrument the employment of which has brought both temporary and permanent relief, sometimes cure, to thousands suffering the most intense agony. Its value in cases especially suited to it cannot be over-estimated; its use in cases unsuited to it, or where other means would answer as well, cannot be too strongly and heartily condemned. Death, syncope, alarming narcotism, and, perhaps, more important than all, that living death— the morphia habit—bids us choose our cases well, and continue its use for only so long a time as is absolutely necessary.

" It would seem, from a study of the cases related under the latter head, and from many of the interesting letters for which I cannot find space, and, therefore, give at second-hand, that so long as the drug is used carefully, and with discrimination, by the physician, the morphia habit is little apt to result, and that it may be broken off when once formed, although the amount of morphia used is large. But, nevertheless, even when these results are attained, persons, especially those of the neurasthenic type, will often procure a hypodermic syringe surreptitiously, and continue, commence or re-commence the practice anew. For this the physician is certainly not to blame, but the fact still stands, and the question arises whether relief by other means, though less prompt and less permanent, would not have been more preferable to the deplorable mental, moral and physical condition that almost uniformly obtains when the habit is once established.

"When Wood first gave his idea to the world, and when that idea was practically developed and extended in point of applicability by Hunter and others, it was thought to be glory enough to have found a weapon sufficiently powerful to cope with severe and obstinate neuralgia and diseases that would not yield to other treatment. To-day, as seen from its journal literature, the subcutaneous method of giving morphine has become almost universal, and it is employed for complaints of the most trivial character. Aside from the immediate and remote danger of thus needlessly extending this practice, there is another reason why this instrument should not be so commonly employed— there is apt to be slurring of diagnosis and a blind treatment of the most prominent symptom. This is especially the case with the younger members of the profession. Having relieved the pain, they fail to study the minor symptoms, to look at patient's family and personal history, to be observing of skin, and tongue, and pulse. It may be urged that the patient often recovers without any other treatment. True! But many do not, and the lack of study of every point in the successful cases bears its fruit of slovenly diagnosis and unscientific treatment in many succeeding cases.

" Finally, let it be distinctly understood that I consider the hypodermic use of morphia a very decided therapeutic advance, and of incalculable benefit in allaying pain and curing disease in cases where other modes of treatment utterly fail. In calling attention to the dangers sometimes attending the use of the drug in this way, I do it not from a one-sided view of the question, not with a desire to condemn it, simply to point out what *may* occur, endeavor to show how best to avoid it, and, if possible, restrict its use to those cases in which it is proper.''

The physician should never entrust a syringe or the solution to patients or their friends. *He should use it himself.*

THE CHLORAL HABIT.

CHAPTER VIII.

THE CONTINUED USE OF CHLORAL HYDRATE.

But little attention has been paid in this country to the habitual use of chloral. German and English literature is more abundant, but at best vague and unsatisfactory, and the professional mind does not seem to be at all settled upon the subject as to whether such a thing as a chloral habit does or does not exist.

A committee was appointed by the Clinical Society of London, a short time ago, to investigate this matter. The result was a failure, owing to the fact that the medical gentlemen who received circulars asking for information upon various points failed to answer in sufficient numbers or with sufficient explicitness to make the results valuable. There were but seventy replies from the thousand circulars sent out.

Not knowing until afterward of the work then undertaken by that society, I prepared and sent out some ten thousand circulars, containing a series of questions upon chloral, one of which related to the habitual use of the drug. To this circular I received over four hundred replies, one hundred and seven of which related to the chloral habit.

149

150   THE CHLORAL HABIT.

One hundred and seven physicians reported 135 cases where the long continued use of the drug had led to a more or less marked craving for it.

An analysis of these cases may prove of interest.

The ages of the patients were obtained in ninety-four instances, and show as follows :—

From 20 to 30 . . .	12
From 30 to 40	25
From 40 to 50	27
From 50 to 60 . . .	19
From 60 to 70 . . .	11
	—
	94

The sex was obtained in one hundred and thirteen cases, as follows :—

Males	77
Females	36
	—
	113

There were among the number clergymen, physicians, editors, clerks, prostitutes, mechanics, farmers and hospital nurses.

The majority of the females were married women, suffering from painful uterine complaints. The majority of the males, persons who had been or were addicted to the use of alcoholic stimulants. Some, both males and females, formed the habit through taking it for insomnia and mental depression, dependent on family troubles, business failures and the like. Some through hearing or reading of its magical effects and trying it to satisfy themselves.

In many important points the habitual or long continued use of chloral differs from that of opium or morphine, viz :

1. Many persons take it for years without any obvious ill effect.

2. Its continued use is not so liable to be followed by a morbid craving for it as for morphine or opium.

3. The habit once formed is, in the majority of instances, easily broken.

4. An abrupt change from chloral to some other narcotic is often made.

As regards the first and second propositions the following facts may be taken into account :—

Dr. W. M. Compton,* Medical Superintendent, Insane Asylum, Jackson, Miss., says that he has given twenty to thirty-grain doses of chloral hydrate nightly, for years, to noisy patients, without observing any ill effects. Dr. J. H. Nordlin,† of Rome, Ga., has given it in combination with bromide of potassium, capsicum and ammonia, in asthma, for a long time, without any ill effects.

A correspondent of the London *Lancet*,‡ a medical man, says that, for sleeplessness, he has been obliged to take twenty-grain doses of chloral at bedtime, for upward of four years, without any bad consequences whatever.

Dr. J. G. Thornley § has used chloral in the treatment of insane persons, continuing its use for from six months to two years, without any ill effects.

Dr. Schlaugenhausen, First Assistant Physician to the Lunatic Asylum, Hall, Tyrol, writes me that in a case of insanity he gave thirty grains of chloral every night, for six years, without any perceptible bad effect. He says that he has never seen a case where there was a morbid craving for chloral, such as is seen from opium or morphine. Dr. John Wirtinger, of Ybbs on the Danube ; Dr. Reiman, of Kiev, Russia, and Dr. Fr. Akland, of Stockholm, Sweden, write the same.

* N. Y. *Medical Record*, 1876, p. 436.   † By letter.
‡ April 14th, 1877.   § *Lancet*, Dec. 18, 1875.

Dr. A. Ady,* of West Liberty, Iowa, gave chloral in a case of chronic diffuse nephritis, for nearly a year and a half, without any bad results and with no tendency to the formation of a habit.

Bidlack† relates a case where this drug was used for a year, with no bad effects.

Lawrence Turnbull‡ gave chloral, in medicinal doses, for one year, as a sedative and narcotic. Its use was stopped at once, without any ill effects whatever. Dr. Blackwood§ used thirty grains three times a day, in combination with bromide of potassium, for three months. No bad effects beyond slight conjunctivitis. No evil effects followed the withdrawal of the drug.

Dr. Carlos F. MacDonald‖ writes: "As regards the so-called ' chloral habit,' I can only say that I have used chloral very largely in my practice—commencing with its introduction to the notice of the profession in America— and I have yet to meet with a case."

A physician, who does not wish his name used, took twenty grains at bedtime, for insomnia, for nearly two years and a half, without ill effect. Professor A. A. Smith, of Bellevue College, this city, writes me: "A lady of an extremely nervous temperament had it administered to her during an attack of typho-malarial fever. This was seven years ago, and she has taken it every night of her life since. Usually ten grains suffices, but sometimes she takes thirty grains. It has had no perceptible bad effect. She seems perfectly well. I have, time and again, tried to wean her from the habit, but to no purpose, and I am quite certain that the drug has never yet done her any harm."

* Letter.
† Philadelphia *Medical and Surgical Reporter*, Nov. 9, 1878.
‡ Ibid.   § Ibid.
‖ *American Journal of Insanity*, 1878, p. 367.

Dr. G. A. Shurtleff, Medical Superintendent State Insane Asylum, Stockton, California, writes me that he has seen a morbid desire for chloral produced by the taking of from twenty to forty grains regularly at bedtime, to procure sleep. It was easily broken, the desire passing away in a few days.

Dr. M. L. Holbrook, of this city, writes me of the case of a highly cultured lady, an authoress, now in Germany, who had taken thirty grains of chloral nearly every night, for eight years, without the production of a morbid craving or other ill effects.

Dr. Joseph Parrish, of Burlington, N. J., writes me of the following curious case:—

"I have known persons (and have one under treatment at this time) who are habituated to the use of narcotics, and who, to avoid being enslaved by either one, alternate between morphia, chloral, whiskey, and the bromides, with the result of rest and composure from each in its turn. They are not opium or chloral habitués, nor inebriates from alcohol, but they are habituated to artificial narcotism.

"In the case now under treatment, the chloral dose does not exceed fifteen grains at bedtime, and the morphia dose does not exceed half a grain, nor the whiskey dose more than a fluid ounce. I learn from the patient that his former physician employed these remedies for long-standing insomnia.

"Present condition fair; i. e., delicate constitution, but good appetite, and even temperament; functions not interfered with, though the habit of taking something to promote rest and sleep is of several years' duration."

The following is from a prominent physician, who does not care to have his name appear in connection with the case:—

"A lady patient of mine—now fifty-seven years of age —began to take fifteen grains of chloral about eight years ago, after very great family affliction, to procure sleep, and has kept up its use almost uninterruptedly ever since, without increasing the dose, though she occasionally (not very often) repeated the dose once during the night. During seven of these eight years she was subjectéd to the depressing influence of the care of a child with severe epilepsy and mental deterioration slowly progressing. During the last year she has gone several times, for a week at a time, without the drug, and I think that she could now dispense with it almost entirely were it not for constant anxiety about one of her children. She takes it only to procure sleep; and I cannot see that it has harmed her in any way.

"The above case cannot be called one of the chloral habit, as during a number of hours out of every twenty-four she is free from the influence of the drug; but I think the case may be of value as illustrating the length of time during which the drug may be taken, almost daily, without any increase of dose."

Professor Henry M. Lyman, of Rush Medical College, Chicago, writes as follows:—

"I am acquainted with a clergyman, sixty-five years of age, whose mother died insane. He informs me that for ten years past he has been obliged, in order to procure sleep, to take ten grains on retiring, and ten grains more about 3 A.M. His health is otherwise good, and he is possessed of more than common intellectual power. He, however, though never in the habit of using alcoholic drinks, has the red eyes and rather 'groggy' appearance of a person who drinks too much. I have known of other cases where persons were in the habit of taking ten grains at bedtime, every night, for months together, with great benefit. I have never happened to make the acquaintance of a person who was addicted to its use as a simple instrument of pleasure.''

Some men, in the early history of this drug, and some at the present day likewise, lauded it highly as a substitute for opium, claiming for it perfect freedom from the danger of forming a habit and the non-production of ill effects when used for a long time.

Thus a correspondent of the *Doctor** says :—

''Long-continued use does not impair the general health. I know persons who have taken it almost since its introduction, in the same dose and with the same success.''

Dr. Pollak† says: ''The prolonged employment of chloral is not usually attended with any disagreeable effects, and if any such occur, they are of no consequence. It especially does not induce congestion of the brain or disturbance of the digestive organs.''

Dr. C. W. Cram,‡ of Columbus, Ohio, writes: ''Its use is not followed by that craving for its continuance which so often attends the administration of preparations of opium, especially morphine, producing a multitude of opium-eaters.''

Others went to directly the opposite extreme.

Dr. Madison Marsh § says: ''Its effects are so pleasant, its use so exquisitely fascinating, that, the habit once acquired, a person becomes a slave to its use, never to stop till death closes the scene. The enchantments of alcoholic stimulants, cannabis Indica, morphine or tobacco, bind with silken cords, compared to the bars and hooks of steel

* New York *Medical Record*, 1872, p. 106.
† *Wiener Med. Wochenschrift*, Feb. 28, 1874.
‡ The *Clinic*, March 2, 1872.
§ Philadelphia *Med. and Surg. Reporter*, Jan. 16, 1875.

thrown around the unhappy victim of this popular drug and infatuating stimulant.''

The first to call attention to the possibility of the formation of a "chloral habit" was Dr. B. W. Richardson, of London, a part of whose most excellent paper* I cannot refrain from giving.

" It is a matter of deep regret to have to report that, since the name was given to the disease, chloralism has become widespread.   It has not yet spread far among the female part of the community.   It has not yet reached the poorer classes of either sex.   Among the men of the middle class, among the most active of these in all its divisions—commercial, literary, medical, philosophical, artistic, clerical—chloralism, varying in intensity of evil, has appeared.   In every one of these classes I have named, and in some others, I have seen the sufferers from it and have heard their testimony in relation to its effects on their organizations.   Effects exceedingly uniform and, as a rule, exceedingly baneful.

" At the meeting of the British Association for the Advancement of Science, held in Edinburgh, in the year 1871, I drew earnest attention to this subject.   I said, and the words were published in the reports of that year (p. 147), there is another subject of public interest connected with the employment of chloral hydrate.   I refer to the increasing habitual use of it as a narcotic.   As there are alcoholic intemperants and opium eaters, so, now, there are those who, beginning to take chloral hydrate to relieve pain or to procure sleep, get into the fixed habit of taking it several times daily, and in full doses.   I would state from this public place, as earnestly and as forcibly as I can, that this growing practice is alike injurious to the

* *Druggists' Circular and Gazette*, Nov., 1870 .

mental, the moral, and the purely physical life, and that the confirmed habit of taking chloral hydrate leads to inevitable and confirmed disease. Under it the digestion, natural tendency to sleep, and natural sleep are impaired ; the blood is changed in quality, its plastic properties and its capacity for oxidation being reduced ; the secretions are depraved, and the nervous system losing its regulating. controlling power, the muscles become unsteady, the heart irregular and intermittent, and the mind excited, uncertain, and unstable. To crown the mischief, in not a few cases, already, the habitual dose has been the last ; involuntary, or rather, unintentional suicide closing the scene. I press these facts on public attention not one moment too soon, and I add to them the further facts that hydrate of chloral is purely and absolutely a medicine, and that whenever its administration is not guided by medical science and experience, it ceases to be a boon and be comes a curse to mankind.

" This was stated within two years after the substance, chloral hydrate, came into medical use. If at that time the mind of the public had been as ripe as it is now for the acceptance of the truth, or if I could then have reached the ear of the public more plainly, much evil might have been nipped in the bud. As it was, the warning had little effect, except to expose me to adverse criticism as an alarmist, and the evil has gone on with increasing rapidity and mischief. There is, at the present time, a considerable community addicted to the habitual use of chloral hydrate, on one pretence or another, and a learned medical society has recently framed a series of written questions on the subject, which questions it has felt it expedient to address to members of the profession of medicine generally, for their replies.

" The persons who become habituated to chloral hydrate
are of two or three classes, as a rule. Some have originally
taken the narcotic to relieve pain, using it, in the earliest
application of it, for a true medicinal and legitimate object,
probably under medical direction. Finding that it gave
relief and repose, they have continued the use of it, and at
last have got so abnormally under its influence that they
cannot get to sleep if they fail to resort to it. A second
class of persons who take to chloral are alcoholic inebriates
who have arrived at that stage of alcoholism where sleep is
always disturbed, and often nearly impossible. These
persons at first wake many times in the night, with coldness
of the lower limbs, cold sweatings, startlings, and restless
dreamings. In a little time they become nervous about
submitting themselves to sleep, and before long habituate
themselves to watchfulness and restlessness, until a confirmed insomnia is the result. Worn out with sleeplessness,
and failing to find any relief that is satisfactory or safe, in
their false friend, alcohol, they turn to chloral, and in it
find for a season the oblivion which they desire, and which
they call rest. It is a kind of rest, and is, no doubt, better
than no rest at all, but it leads to the unhealthy state we
are now conversant with, and it rather promotes than
destroys the craving for alcohol. In short, the man who
takes to chloral after alcohol enlists two cravings for a
single craving, and is double shotted in the worst sense.

" A third class of men who become habituated to the
use of chloral are men of extremely nervous and excitable
temperament, who by nature, and often by labors in which
they are occupied, become bad sleepers. A little thing in
the course of their daily routine oppresses them. What to
other men is passing annoyance, thrown off with the next
step, is to these men a worry and anxiety of hours. They

are over susceptible of what is said of them and of their work, however good the work may be. They are too elated when praised, or too depressed when not praised or dispraised. They fail to play character parts on the stage of this world, and as they lie down to rest they take all their cares and anxieties into bed with them, in the liveliest state of perturbation. Unable in this condition to sleep, and not knowing a more natural remedy, they resort to the use of such an instrument as chloral hydrate. They begin with a moderate dose, as occasion seems to demand, and at last, in what they consider a safe and moderate system of employing it, they depend on the narcotic for their falsified repose. Among these classes of men the use of chloral hydrate is on the increase. The use is essentially a bad business at the best, and while I do not wish in the least to exaggerate the danger springing from it, while, indeed, I am willing to state that I have never been able to trace out a series of fatal organic changes of a structural character from such use, I have certainly seen a great deal of temporary disturbance and enfeeblement from it, without any corresponding advantage that might be set forth as an exchange of some good for some harm. The conclusion I have been forced to arrive at is, in brief, to this effect, that if chloral hydrate cannot be kept for use within its legitimate sphere, as a medicine to be prescribed by the physician, according to his judgment, and by him as rarely as is possible, it were better for mankind not to have it at any price.''

A case showing the evil effects of the continued use of chloral in large doses, and aptly illustrating how readily some of these patients pass from the use of this drug to alcoholic stimulants and *vice versa*, is related by Dr. T. Inglis,* of the Royal Edinburgh Asylum. I here give an abstract of his very interesting article.

* Edinburgh *Medical Journal*, Sept. 1877, p. 211.

"F. S. P., shopkeeper, male, aged forty-seven. Patient's father died of kidney disease, at an advanced age. Mother "nervous," died of paralysis. Two sisters neurotic and eccentric. Brother a confirmed dipsomaniac, died of brain softening. Patient a man of average intelligence, neurotic temperament, previous general good health, but has had asthma. Habits teetotal for ten years, up to three weeks before admission.

" Seven years before was ordered chloral and bromide of potassium, for relief of spasmodic retention of urine. Of these he took about one drachm each, daily, for six years regularly, and during that time neither he nor his friends noticed any ill effects on either mind or body.

" At the end of this time he had an attack of bronchitis, for the dyspnœa attending which he was given chloral, without bromide. Recovery from bronchitis was rapid. Death in his family and other troubles, however, led him to the continued use of the chloral. Commencing with sixty grains daily, he came to use one hundred and eighty grains daily. This in ten-grain doses, often repeated. Effect not hypnotic, but calmative and soothing.

" No headache or vertigo were complained of, but there was lassitude, nervous debility and exhaustion ; irritability of temper and peevishness. Anorexia, acidity, nausea and vomiting appeared. Piles became troublesome and the fæces white and hard.

"Intellectual enfeeblement and perversion of affective life and character. He became untruthful and deceitful, and manifested a dislike for home and family. Even threatened violence to his wife. Three weeks before admission he discontinued using chloral, and kept himself in a chronic state of muddle and confusion, by means of whiskey. In a day or two he became restless, excited and quite unmanageable. Diarrhœa set in, with great discharge of blood from the bowels. A state resembling delirium tremens came on. Imaginary voices mocked him; he saw loathsome animals, snakes, etc. What little sleep he had was broken by terrible dreams. This terminated in three severe and unmistakable epileptiform convulsions, at intervals of four hours. This was followed by stupor, and that by excitement.

ON ADMISSION.

"Man of average height and development, but seems to be prematurely old and broken down. Very weak and anæmic, can speak only in an undertone, and can hardly walk.

"Expression blank and vacant. Eyes dull and meaningless. Great enfeeblement of mind. Silly, childish, and almost imbecile. No excitement, but, rather, slight depression of mind. Would laugh and cry without adequate cause. No interest in surroundings. Replies to questions rambling, incoherent and disconnected, being unable to carry on a consecutive line of thought. Memory almost obliterated. Could not tell his age or where he came from. Had vague, fleeting delusions on various subjects.

"Persistent muscular tremulousness of upper and lower extremities, requiring assistance in walking, while the finer acts of coördinative power, as writing and whistling, could not be performed at all. Tongue furred in the centre, tremulous and with fibrillary twitching at the edges, and pointed markedly to the right side. Articulation thick and indistinct. Pupils equal, dilated, irregular at margins, and insensible to light. Right side of face partially paralyzed. Reflex action of cord much impaired. Common sensation acute, verging on hyperæsthesia. Complained of sleeplessness and exhaustion, but had no headache nor neuralgic pains.

"No cutaneous eruption. Muscles poor and flabby. Conjunctivæ yellow. Respiratory and circulatory system normal. Pulse 67, weak and thready. Temperature 97°. Urine, specific gravity, 1.005. Neither albumen, sugar, bile nor tube casts. Could only make water at night, when it was passed in large quantities, was clear and limpid, resembling that of hysteria. Bowels confined, fæces hard and of a white color.

" Neither narcotics nor alcoholic stimulants were used. Tonic mixture, containing strychnia, prescribed. Improvement in all the symptoms marked and rapid. Discharged, well, physically and mentally, in three months' time."

This case illustrates, also, the fact that the majority of those who become habituated to chloral, and in whom it works the most serious injury, are of the neurasthenic type, and manifest a morbid craving for some stimulus or narcotic.

Of the cases where I have been able to obtain information upon this point, I find the temperaments given as follows:—

Nervous,	63
Sanguine,	1
Nervo-sanguine,	8
Melancholic,	21
	93

In forty-one cases there is a history of either nervousness, nervous disease, insanity, alcoholism or opium addiction in the ancestors.

In many of these patients there is a "mixed" habit, as

opium and chloral, alcohol and chloral, alcohol, chloral and morphine, chloral and chlorodyne, etc.

It is largely used* by the intemperate as a substitute for alcohol. On this point the testimony is abundant and positive.

As an example of hereditary tendency, etc., the following, sent me by Dr. N. Jasper Jones, of Blairstown, Iowa, is very interesting:—

"Alcoholism prevalent in the family, also the opium habit, but no insanity, although the mother, at times, under mental worry, becomes somewhat unreasonable, as though she might become insane. Father died of heart trouble while sitting at breakfast. The entire family are of the most intense nervous type, on the mother's side, all the children taking after the mother; but the father was unusually reserved and cautious, probably because he had schooled himself to be so, on account of his wife's and children's excitable natures."

It will be seen, from the foregoing cases, that while there is no doubt that in certain individuals a morbid craving for chloral may be established by the prolonged use of the drug, it is very different, as a habit, from that produced by opium or morphine. As an offset to this stands the fact that, in those cases where a habit is formed, *it causes a more complete and rapid ruin of mind and body* than either opium or morphine.

Persons addicted to the habitual use of this drug are usually of the educated class, as is the case with opium.

It has been denied by some authors that it is ever used as a stimulant to the mind. In one case now under my care, and in several reported to me by correspondents, this was the object for which it was first taken, and it was

* Dr. A. P. Hayne, Med. Supt. Inebriate Asylum, San Francisco, Cal.

found to answer the purpose admirably, especially in the case of those doing continuous literary work.   This effect was, however, short-lived and the reaction decided.

Dr. Quintius C. Smith, who has had a large professional experience among the prostitutes of the Pacific coast, writes me that the drug is or was used largely by this class, they finding it quite as efficient a stimulant, nerve sedative and sleep producer, as opium or alcohol, and having the advantage of leaving no ill after effects.   One such patient, unless she previously took a dose of chloral, would have a severe epileptiform convulsion at each connection.

A clearer idea of the ill effects of the long-continued use of chloral in large doses is best studied on each system or apparatus separately.   Unlike morphine, it not only shows its baneful effects on the nervous system, but it acts to directly undermine the health by disorganizing the blood.

CHAPTER IX.

EFFECTS OF CHLORAL ON THE DIFFERENT SYSTEMS AND APPARATUSES.

As I said in my last chapter, it is not every person who uses chloral for a lengthened period who becomes an habitué, and it is not every habitué who is affected in every way as fully or completely as in the manner about to be described. In each individual the organs chiefly affected and the length of time necessary to produce the effect vary according to idiosyncrasy, which simply means peculiar personal susceptibility. This same fact holds good with reference to opium and morphine, as well as chloral.

MIND.

The effect of the long continued use of this drug on the mind is, in the majority of instances, much more rapid and distinct than that of opium. At first there is a transient period of stimulation, followed by gradual weakness and cloudiness of intellect, mental hebetude, apathy to what is going on around. A low species of cunning is usually developed, that leads to the planning of plausible excuse and subterfuge to obtain a full supply of the drug. This condition is well illustrated by the following :—

Dr. Wm. Kirkpatrick Murphy* reports three cases of chloralism —

1. Lady, fifty-eight; spare but muscular, and of more than ordinary intelligence and strength of will. Took morphine for some time, to overcome sleeplessness due to painful

* London *Lancet*, Aug. 2d, 1873.

affection of the bladder. Six months before Dr. M. saw her she had been taking chloral, twenty grains at night, to commence, finally one hundred and fifty grains in the twenty-four hours. At his first visit found her almost a skeleton, hard to rouse, maudlin and semi-narcotized. When questioned, answered wildly and disjointedly. State of mind closely bordering on imbecility. Pupils dilated. Pulse slow, feeble and intermittent. Face flushed and anxious. This had been her condition for many weeks. Her rare lucid intervals were spent in childish whining for more chloral, or in cunning attempts to deceive her attendants and obtain an extra dose surreptitiously. From having been a cheerful, upright, exceptionally intelligent and strong-willed woman, she had become morose, deceitful, and imbecile alike in intellect, memory and will.

Points of exceptional interest.

1. Marked decubitus, with not mere indisposition, but, at times, manifest incapacity for exertion.

2. Dark, erythematous flush over head and neck, with extraordinary twitching of facial muscles.

3. Partial paralysis of the œsophagus, not spasmodic stricture ; but sluggish and incomplete contraction of the muscles of deglutition, with regard to both food and drink. This was well marked in Case 2, and in both recurred at intervals for many months ; even after convalescence was established.

4. Dimness of sight ; eyes bloodshot and constantly watering. Once, after total cessation of chloral, temporary loss of vision, lasting, with intervals, two days.

5. Dryness of the hair.

CASE 2.—Lady, aged forty-five, good health and keen intellect. Insomnia from worry and from dyspepsia. Had taken chloral for two years. At first only at bedtime ;

after a time during the day also. Never above seventy grains in twenty-four hours. Both physical and mental exhaustion, morbid fear, confusion of mind, lack of concentration and impairment of memory. (She was, however, an authoress, apparently doing her regular literary work. —K.)

POINTS OF INTEREST.

1. Erythema of head and neck.

2. Excessive hyperæmia after smallest quantity of wine or beer.

3. Partial paralysis of the œsophagus, with nervous disinclination to take food or drink.

4. Dimness of sight, with eyes bloodshot and overflowing with tears. Marked projection of eyeballs.

5. Intense nausea, oftentimes, after taking chloral.

Intense melancholia, with suicidal tendencies, hallucinations and delusions, delirium. Acute and chronic mania, have resulted. A condition, seemingly a cross between idiocy and softening of the brain, sometimes occurs, the patients crying at the most trivial things, and bemoaning their persecution by what would to others appear to be but trifles. The moral nature is profoundly affected, but the tendency to lie does not seem to be so distinct as in the opium habit.

Delirium is not so often found while the patient is taking, as when he stops the drug. In some cases it does not even occur then, but when so happening, is usually very violent.

Convulsive seizures are rare, but do occur, as has been shown by the labors of the Clinical Society, of London. Two of my correspondents speak of such cases. In both, the character of the attack was epileptiform, and occurred in persons who had used large doses of the drug for a long time.

A mind greatly enfeebled by the continued use of chloral is sometimes thrown completely from its balance by the effects of an overdose, or the unusually intense action of an ordinary dose, an accident that is liable to take place at any time. Such a case is related by Dr. Willis P. King,* of Sedalia, Mo., as follows :—

"Mrs. S., aged forty years, first began to use chloral during an attack of rheumatism. Has now been taking it for some years. Dose about 32 grains, taken from three to five times a day. I was treating her for acute rheumatism, and allowed her, by her request, to take an occasional dose of a solution that she had been using for years. She begged for more than I was willing to allow, and, in the absence of the nurse, got up, and, from the amount missing, must have taken anywhere from 100 to 200 grains. I found her, three hours afterwards, in profound stupor, pulse small and 190 per minute, respiration rapid and stertorous, extremities cold. Under stimulation, etc., she recovered consciousness after a few hours, but was insane, and was sent to the Missouri Insane Asylum, where she was treated for acute mania. She was always 'flighty' before she went to the asylum; on her return from there her mental condition was greatly improved."

The temper becomes irritable, peevishness is developed, friends and acquaintances are shunned, and the patient lives in a world of his own, built by the warped and distorted outcome of a diseased imagination. The delusions and hallucinations are more often terrifying than pleasant, varying, however, somewhat, with the mental peculiarities

* Letter to the Author.

of the victim. The following interesting case is sent me by Dr. D. N. Kinsman, of Columbus, Ohio :—

"Patient's age, twenty-four years. Had been surrounded by a set of vicious companions, and got to drinking excessively. Had an attack of delirium tremens, for which chloral, bromide of potassium and valerian had each been prescribed. In a fit of virtuous indignation at his oft-repeated falls, he signed a 'Murphy pledge.' Having learned the value of chloral for the horrors, now, in his abstinence from alcohol, he began the use of chloral. He would sit in the office and sleep in his chair, day after day. His speech became drawling and indistinct, and he could not remember what he read. At the end of two or three weeks he had an eruption appear on his face, neck, and shoulders, like acne. These pimples, at first hard and red, suppurated, being attended with much soreness and pain.

"After being in this condition for some weeks he began to hear voices at night, rose from his bed in great terror, and ran to his father's room, saying burglars were in the house, on several occasions. His neighbors were in a conspiracy against him; living in a double house, could hear them planning against him, through the walls. On going home at night would go at full run, waiting till very late, to have the streets deserted; asserted he was often pursued by a 'band of niggers' who had been hired to injure him. Heard people talking about him, whom he passed, even when I was with him and knew nothing was said. He continued in this state for several weeks; finally became exhausted, had to take to his bed, where he remained several weeks. He had been removed to his father's farm, several miles distant, at this time, hence, not at this period under my observation. Now, being obliged to abstain, he got better; but when he got better, relapsed into his old habit. Now, he was sent to Fort Hamilton, and there kept for, I think, six months. Perhaps a record of this case can be found there.

"He returned here, and has been fully restored. Though after his sojourn in Fort Hamilton he had another attack of his appetite for chloral, whiskey, etc., which lasted some months.

"His doses of chloral, daily, would amount to half an ounce, I estimate, for days together. The data are impossible to learn, because he never kept it all in one bottle, and he had a supply located at various points, for convenience.

"The eruption did not itch. It lasted during the time of taking the drug, and subsided after it was discontinued. There was also, at times, a use of valerian, in quantities unknown."

Dr. Roswell Park, of Chicago, Ill., writes me of a young man "who established the chloral habit after a series of business reverses. He was nervous and excitable. I have no idea as to what amount he took. He finally developed a suicidal tendency, jumped from a fourth-story window, sustained a compound comminuted fracture of the elbow, and other injuries, and finally died, about the tenth day, from acute mania and exhaustion."

Dr. James Bunting* had a patient who took large doses of chloral daily, with the effect of perverting the morals and producing a tendency to suicide.

Dr. Thos. Bond† relates the case of a friend, a young medical man, who died of acute mania, produced by continued and excessive use of chloral.

The memory, especially for words, is greatly impaired.

*Lancet, Dec., 1875.   † Lancet, March 11, 1876.

I have seen one case of this kind, and one such was reported to me by a medical gentleman in Boston, who forgot to sign his name to his very interesting letter.

The patient whom I saw not only had an impaired memory for words, but would use one word for another, as "slippers" for "hat," or "pen" for "coat." He had been using the drug for nearly six years, and, although profoundly affected by it, would, at times, be decidedly better than at others.

Dr. G. W. Davis, of Chico, California, writes me of the case of a lady who had been using the drug for some time. She first took it for facial neuralgia. She takes from forty to sixty grains daily. Her present health is fair. "She complains, however, of a fullness about the head, confusion of ideas, and loss of memory."

Drs. Norton Folsom, of Boston, and J. W. Parsons,* of Portsmouth, N. H., relate cases where the habitual use of this drug produced childishness, fretfulness, muscular tremor, and, in Dr. Parsons' case, seemed to aggravate an existing attack of melancholia.

Dr. James Perrigo,* of Montreal, relates the case of an old lady, sixty years of age, who took chloral, forty-five grains at bedtime, and fifteen grains during the day. She was partially insane, but whether from the chloral or not, I have been unable to learn.

Maudsley* uses it but rarely in insanity. "Sometimes in melancholia. Rarely in mania, and then only when subacute." Similar testimony was obtained from Drs. Maudsley, Clouston and Lindsay, by the Committee of the London Clinical Society.

Dr. A. P. Hayne, Medical Superintendent Inebriate Asylum, San Francisco, Cal., writes me : " There is no

* Letters to the Author.

doubt but that its long continued use is one of the causes of insanity. I think I may safely say that several cases of this kind have fallen under my immediate observation, where chloral has been the remote or immediate cause of the disease.''

Dr. H. H. Doane,* of Litchfield, Ohio, reports a case where chloral produced, in an habituè, melancholia, with suicidal tendency.

The following interesting case history was sent me by Dr. Frank R. Fry, of St. Louis, Mo. :—

'' Patient, J. M., aged fifty-nine years. First used chloral eleven years ago. At that time he had been having some domestic troubles, had been drinking some—he says, not hard ; he was unable to sleep, on account of these troubles and severe pains in the soles of his feet. His physician said they were not caused by rheumatism, but did not say by what. His physician prescribed him 15 grains of chloral in one ounce of water. Two teaspoonfuls of this solution, the prescribed dose, gave the patient some relief at first, but he immediately increased the dose to four teaspoonfuls. Does not remember exactly how soon, but very soon, he was taking from thirty-five to forty grains at a dose, every night, to obtain sleep. He increased the dose as fast as was necessary to obtain the continued desired effect, until he was taking eighty grains every night, and often a large dose during the daytime.

Six years ago he was in the City Insane Asylum, for three months. When he left there he felt very much better, and states positively that he used no chloral for a period of three years after that time.

He began the use of chloral again, subsequently, and for the last two and a half or three years he has been using it

* Letter to the Author.

very steadily and in large doses. He does not take such large doses at present, but takes it much oftener than he did. Says the size and frequency of the dose are much of the time varied according to his means. He does not now pay any attention to the number of grains he is taking, as he used to, but he has a hard time, often, in getting enough to satisfy. He is also drinking to some extent now. Says he has tried to substitute alcohol and morphine for chloral, but has always fallen back to the latter.

He is now in bad health, suffers from dyspepsia and constipated bowels, and a generally debilitated condition. His father was a hard drinking man and his mother a very " nervous" person.

Dr. C. Pollock,* of Donnelsville, Ohio, reports the case of a physician, aged fifty. From twenty to sixty grains were used nightly for two years, with the result of producing loss of co-ordinating power, marked failure of memory, loquacity and intense despondency, and obstinate insomnia, only relieved by continued taking of the drug that had caused all his misfortune, and which brought on only a troubled slumber, laden with dreams horrible beyond description. The most prominent features of the case were his profound melancholia and extreme loquacity.

A somewhat similar case is reported by Dr. T. D. Crothers,† Superintendent Walnut Hill Inebriate Asylum, Hartford, Conn., the case ending fatally.

As in the opium and morphine habits, only more so, business is abandoned, friendship broken, family ties sundered, and unless relieved the victim sinks into a state of slobbering insanity or acute mania, finally ending in death through hemorrhage, exhaustion, or from the satisfaction of a sui-

* Quoted by Mattison, " Chloral Inebriety," a paper read before the King's Co. Medical Society, April 15th, 1879.
† Mattison, op. cit.

cidal tendency. Physical wrecks, guided by shattered mental rudders, they sink out of sight or go to pieces through accidents incident to their own self-produced restlessness.

Thirty-eight of my correspondents report the production of insanity by the habitual use of this drug, as follows :—

Acute mania, ...	8
Melancholia, ...	16
" Insanity,"	3
Silliness, childishness, etc., . .	11
	38

In many cases where I wrote for more full and positive information I received no reply.

THE NERVOUS SYSTEM.

On the nervous system the effect of chloral, when habitually used, is quite as decided and injurious as upon the mind.

These symptoms may be classed as :—

(*a*) Those affecting the nerves themselves.

(*b*) Those affecting other organs, or systems, through the medium of the nervous apparatus supplying them.

To the first class belong anæsthesia and hyperæsthesia, usually of *parts* of the body only. These may and do extend to the mucous, as well as to the cutaneous surfaces, as evidenced by gastralgia, loss of taste, incontinence and irritability of the bladder and sexual organs, etc., to be more fully considered a little further on.

Neuralgic pains are felt here and there, more especially in the neighborhood of, *but not in*, the joints. Mattison* and others lay great stress upon this pain, which usually *girdles* the limb above or below the joint, as a diagnostic mark of the chloral habit. That it is *not* distinctive of

* Op. cit., p. 5.

this affection, and occurs in only a few cases, is proven by the fact that out of all the cases I have been able to collect, this symptom appears in but a few—nine—and in not all of these is the pain in the neighborhood of a joint. I have taken special pains to investigate this matter, for the value of a positive diagnostic sign in these affections cannot be underrated. In many of the cases, pains, resembling those of rheumatism or neuralgia, were present, but they did not girdle the limb or occur in the neighborhood of joints.

Dr. Mattison says that Dr. Lee and others have observed these pains in the wrists and elsewhere. I find a case reported by Dr. Benj. Lee,* of a very nervous woman, who had taken chloral in twenty-grain doses, nightly, for eighteen months. Conjunctivitis occurred, the dose was reduced, and the eye trouble disappeared. She again increased the dose, when pains appeared in the wrists, *running along the course of the radial arteries.* The chloral was stopped and the pains disappeared. I find no reference, in the place mentioned, to pain *girdling* the limbs. Unfortunately, Dr. Mattison makes no reference to the source of his information, so that it is impossible to say whether this is the Dr. Lee referred to or not.† Anstie, as quoted by Dr. Mattison, said that he was of the opinion that many cases of supposed rheumatic or neuralgic pains would be found, on careful investigation, to be due to chloral taking, and cited the following case, in which this symptom was prominent:—

"A. B., physician, began the use of chloral Feb. 1, 1873, in thirty-grain doses, to produce sleep, when kept

* Philadelphia *Medical and Surgical Reporter*, Nov. 9, 1878.

† In a letter just received Dr. Mattison states that Dr. Benj. Lee, of Philadelphia, is the gentleman referred to, but that he (Dr. Mattison) does not remember in what journal he saw the case reported.

awake by great anxieties. In two months noticed inflamed and weakened eyes, with scalding tears. Continued
the drug, however, sometimes increasing the dose and
repeating it. From April to August the usual amount was
one drachm. In this latter month commenced using it
during the day, one to three times. About Dec. 1st, began
to realize the amount he was daily taking, and found it
half an ounce, sometimes more. He now began to complain of severe general pains, especially about the joints,
which grew worse in the moist air of London. There was
no tenderness, and they were not increased by motion.
Chloral did not relieve them, except when it put him to
sleep. Soon after this he made a mistake in his dose,
using from a strong solution, which brought on the pains
with frightful severity, and Dr. Anstie was summoned.
He found him with suffused eyes, haggard features, sleepless, peculiar, broken speech, lower extremities partially
paralyzed, with loss of coördinating power, and excessive
joint pains. An examination disclosed that he had taken
more than an ounce of chloral the preceding day. It was
at once withdrawn, cannabis Indica used to relieve the
nervous disturbance, tonics ordered, and under this treatment he recovered.''

It is a well-known fact that rheumatism or rheumatic
neuralgia, so called, is often aggravated by the administration of acids. It is barely possible that in this case, the
drug being kept in a solution, probably with syrup,
partial decomposition may have taken place, and through
the large amount of chloral taken, a sufficient amount of
hydrochloric acid has been ingested to produce the
aggravation of pain. I offer this simply as a suggestion.
Many chloral-takers, especially the English, use the drug
in the form of the officinal syrup chloralis (B. P.), or in

a syrupy solution, expose it to the air and light, and do not keep it in warm places. When a large amount of the drug is used, and decomposition has taken place, an appreciable amount of acid might be taken.

Pains in the limbs, face, chest, eyes, and about the heart are not uncommon, but in some cases never occur, although the person is taking large amounts (100 to 200 grains).

Flushing of the face and ears, and congestion of the eyes, the latter apparently bloodshot, is often seen, and is intensified by the use of a small amount of any alcoholic stimulant. A physician in the South, an habituè for years, writes me that his face became so very scarlet, and his eyes so bloodshot after drinking a glass of liquor, that people in the street would turn around and stare at him in amazement. It is of the same character, though less intense than that flushing of the face produced in some people, not chloral-takers, by the use of coffee. It is due to a more or less complete paralysis of the vaso-motor nerves. It is usually accompanied by dyspnœa and palpitation of the heart. It sometimes occurs on taking stimulants a week or ten days after the habit has been broken.

Nervous chills are also often found, and following them is sometimes felt numbness of one leg or arm, or the sensation of insects crawling on the skin of the trunk, less often of the limbs.

Paralysis sometimes occurs, and is usually confined to the legs. Dr. W. R. Upham, of Yonkers, N. Y., reports the following :—

"Had one case of a lady afflicted with cancer of the breast, who had taken about twenty-five to thirty grains, for one year. Soon after this she became partially paralyzed in the lower extremities; and, becoming very much frightened, she discontinued the use of the drug at once,

12

and without much trouble. She afterwards took up the use of morphine, and has recently died from the original disease. She was of a very nervous temperament."

Here is a similar case by Dr. William Kirkpatrick Murphy*:—

"Young man, twenty-nine years of age. Muscular athlete. Took chloral for insomnia. Amount taken not known, as it was taken loosely, without measure. Once nearly died from an overdose. Awoke one morning to find the power of his lower limbs completely gone. This was transient, but caused him to abandon the use of the drug.

POINTS OF INTEREST.

"1. The necessity often to use a large amount, in order to obtain any effect. When awakening, thirst, nausea, headache, irritability.

" 2. Well-marked erythema, intensified by stimulants. Palpitation very severe.

" 3. Acute pains in lower limbs, especially calves of legs. Some loss and perversion of sensibility in feet and legs in morning.

" 4₊ Inability to use legs. Passed off in one day, leaving patient physically and mentally weak."

Brown-Sequard† relates "two cases of mania, with obstinate insomnia, in which he gave five grains twice a day to one and double that to the other, followed by thirty to forty grains at night. After seven or eight weeks of this treatment the patients had become so feeble as to be unable to walk or to put one foot before the other. This rapidly disappeared when the chloral was withdrawn."

Anstie (also quoted by Dr. Mattison), with his characteristic honesty, relates the following:—

---

* London *Lancet*, Aug. 9, 1873.　　　† Quoted by Mattison, op. cit., p. 5.

" He was called in consultation, to a lady, aged thirty-eight years, with symptoms of partial paraplegia, of obscure origin. Reflex uterine irritation was mentioned as a cause, but the suggestion was not accepted. The regular medical attendant then suggested that the nightly doses of chloral, which the patient had for some time been taking, to procure sleep and allay mental anxiety, might be a factor in the paralytic trouble. This was also rejected, and, as some sedative seemed called for, the chloral was continued. The lady, however, failed to improve from the treatment recommended, and, after one or two consultations, a different physician was called. He accepted the chloral hypothesis, the sleeping draught was discontinued, and she speedily lost her paraplegia. Anstie observed that the case afforded him some practical but rather rueful reflections."

Dr. Mattison says : " The loss of power in the lower extremities is sometimes very marked, and strongly suggestive of serious spinal lesion." It would seem, from the rapidity with which the trouble disappears on the discontinuance of the drug, and its sudden appearance in some cases, to be due, more likely, to derangement of the circulation in the cord, dependent on vaso-motor paralysis.

Vertigo, trembling of the hands and body, and "inability to stand erect, with tendency to fall forward, as in ataxic trouble, lack of coördinating power, so as to be unable to write, whistle, button clothing, etc." (Mattison), are often seen. The tongue, when protruded, trembles ; paralysis of one side of the face and ptosis sometimes occur. Flashes of heat, followed by a cold sweat, on stumbling, "stubbing" the toe, or without any appreciable cause, is often noticeable.

CHAPTER X.

EFFECTS OF CHLORAL ON THE DIFFERENT SYSTEMS AND APPARATUSES.

THE DIGESTIVE APPARATUS.

The effect of chloral upon the organs of digestion is twofold. (*a*). That due to its local action, and (*b*) its effects produced through the medium of the nerves supplying the parts. Taking the throat first, we find that there is usually more or less irritation and congestion from the local action of the drug. In some cases there is paralysis of the muscles of deglutition. Dr. J. H. Arton, of Hamilton, Bermuda, writes me that in one case chloral, *no matter how given*, always produced irritation of the throat and difficulty in swallowing. No other drugs were being given at that time. It disappeared on stopping the chloral. This patient was not an habitual user of the drug. It only produced this effect when the drug was given in large quantity, or often.

Dr. R. C. Brandeis* relates the case of a lady who was an habitual user of chloral. For some nine months she had been taking from forty to sixty grains, three or four times a day. It produced great difficulty in swallowing, intense hyperæmia of the pharynx and larynx, and the return of fluid through the nostrils. There was occlusion of the œsophagus and paralysis of the muscles of the larynx. She fully recovered on strychnia, iron and quinine, finally stopping the chloral. The curious fact has been noted by some of my correspondents, that a

* *American Practitioner*, 1875.

rectal injection of chloral will, in some patients, immediately produce a peculiar sensation in the throat and a metallic taste in the mouth.

" Dyspepsia " is one of the earliest symptoms complained of by these patients. From simple loss of appetite to the gravest forms of gastric trouble may be produced. There is loss of sensitiveness, alternating with the most severe gastric pain. The tongue is coated, the breath foul, and *not* having the odor of chloroform. Circulation in the liver is disordered, gastro-duodenal catarrh results, with accompanying jaundice, that is sometimes intense. A sense of fullness and pain over the hepatic region is not uncommon. Small, brownish patches upon the skin, known to the laity as " liver spots," are frequently seen. The bowels are constipated, the stools lacking in bile, and, consequently, clay colored, hard, and sometimes coated with mucus, here and there streaked with blood. Hemorrhage from the bowels and stomach sometimes occurs. Diarrhœa alternates with constipation.

Some patients, taking the drug in large amounts and for a considerable time, are not troubled at all with affections of the digestive organs.

An habituè, a physician in the South, assured me that it always increased his appetite when taken before meals, and Dr. A. P. Hayne, of San Francisco, says that he has seen a number of these cases, in which there was no effect on these organs beyond a temporary stimulation.

THE URINARY ORGANS.

Upon the kidneys the effect of the long continued use of chloral is to produce congestion, dilatation of the vessels, and albuminuria. There is no positive evidence that chloral produces any organic change in these organs.

Albumen is present in the urine, often in large amount, but its presence there is readily accounted for by the vaso-motor paralysis, and the impoverished condition of the blood, which is still further aggravated by this, its symptom. Casts are often found, but are usually small and medium sized light granular, and hyaline. Epithelial and fatty casts occasionally occur, but it is probable that, in these cases, renal disease had commenced before the chloral was used to excess. Hemorrhage from the kidneys and, indeed, from the whole urinary tract, is sometimes found. It is but a part of a general condition, usually, purpuric spots being found on the skin, and bleeding oc-curring from the mucous membrane of the throat, stomach, nose, or bowels.

Both before the albuminuria appears, and during its continuance, dropsy is not uncommon. Puffiness of the face and eyelids, dropsy of the feet and legs, and ascites occur in certain cases, not from suppression of urine, but from the condition of the vascular system and blood.

The albuminuria, unlike that of opium patients, is usually persistent.

Diabetes is found in some cases where large amounts of the drug have been used for a long time. In acute poison-ing by this drug sugar has been found in the urine of man, by Bouchut* and Levenstein†. Labbée‡ failed to find it in the urine of chloralized rabbits and frogs, but Feltz and Ritter§ found it in the urine of dogs into whose veins they had injected chloral. They proved its presence by both the fermentation and reduction tests. They also claim to have found an organic body, which was present in such

---

* Quoted by Labbée, *Archiv. Gen. de Med.*, 1870, p. 330.
† Berlin *Klin. Wochen.*, July 3, 1876, p. 389.
‡ *Archiv. Gen. de Med.*, 1870. T. 2, p. 330.
§ *Compte Rendu*, Aug. 3, 1874.

small quantity that a determinative analysis was impossible. The same substance was probably found by Von Mering and Musculus, who named it *uro-chloralic acid*. They found that it reduced the copper and bismuth tests, and turned the polarization apparatus to the *left*. This from small doses, while large doses gave sugar, turning the polarization apparatus to the *right*.

Dr. John B. Roberts and Morris J. Lewis* found that the urine of patients taking chloral, and urine to which chloral had been added, gave a yellow precipitate, with Fehling's test solution.

It is probable that in the urine of patients habitually using small doses of chloral, the reaction with the copper solution is due to the presence of uro-chloralic acid, while that of those taking large doses is due to both uro-chloralic acid and sugar; the preponderance of the latter masking or overcoming the reaction of the former, especially with the polarization test.

The functional activity of the kidneys varies greatly from day to day. At times the urine is very scanty and high colored, at others, passed in quantity and of a light straw color. Unlike the urine of opium takers, the gravity varies markedly from day to day (1.006 to 1.045).

On the mucous surface of the bladder and urethra the continued use of this drug sometimes produces decided irritation, and often congestion. Frequent desire to urinate, burning and cutting pain in the act, and a sensation as if the contents of the bladder had not been fully evacuated, are often found, more especially in women.

It has an action, also, on the muscular structure of the viscus, producing temporary paralysis, with retention.

Oxalate of lime, and numerous fibrillæ of mucus, occur as

* *American Journal of the Med. Sciences*, Oct., 1877.

deposits.   The reaction varies from intensely acid to distinctly alkaline.

Deposits of stellar phosphates are of frequent occurrence.
My correspondents report as follows:—

Irritability of the bladder, with tenesmus, 21 instances.
Hemorrhage from the urinary organs,   6   "
Retention of urine, ...   11   "
"Atony" and paralysis,   .   .   .   7   "
Albuminuria,   .   ,   .   .   .   23   "
Reduction of copper test,   .   16   "
Casts in urine, ...   7   "

SEXUAL ORGANS.

The evidence as to the effect of the continued use of
chloral on the sexual apparatus is meagre and conflicting.
The sexual appetite seems to be at first decidedly
increased, but finally greatly impaired or lost.   Such an
increase was observed in one case by Dr. J. A. Miller, of
Williamsburg, Ky. (non-habitué).   I have seen a decided
increase in two cases from occasional small doses, where
the effect could have been due to nothing but the chloral;
and in one case, by Dr. John N. Upshur,* where it had
been used for weeks, with large doses of the bromides, in
a case of tetanus in a negro.

On the other hand, Dr. Edward Bradley, of this city,
writes me that in ten cases he has found decided decrease
in sexual appetite and inability to have an erection in
patients taking. chloral for a short time.   It disappeared
shortly after stopping the chloral.   Dr. J. W. F. Webb,†
of Liberty, Miss., has known it to diminish and temporarily destroy the sexual appetite, when taken only for a
few days.   Also in the case of a chloral eater.

Dr. Horatio C. Bigelow,† of Washington, D. C., saw

* *Virginia Medical Monthly*, May, 1880.   † Letter to the Author.

"loss of erectile power follow the use for one month of ten grains of the drug, three times daily." The person was a strong, healthy man, seemingly of good virile power, to whom it had been given for chronic dysentery. He was never so affected before. It was some months before there was any return of power.

Dr. J. H. Nordlin,* of Rome, Ga., has used it successfully as an anaphrodisiac, as also a correspondent of the *Lancet.*† Dr. O. F. Ham, of North Barnstead, N. H., writes me of a chloral taker, a lady, whom he has questioned closely. She avers that chloral has absolutely no effect on her sexual appetite, which was never marked.

Sensitiveness of the organs of generation, a kind of mucous membrane hyperæsthesia, has also been noted by Dr. H. H. Doane,* of Litchfield, Ohio, as follows:—

"Female, aged forty-two ; used chloral for twelve years. Given for insomnia and nervousness of prolapsus uteri, with constipation. Bilious temperament. Dose, twenty grains, four or five times a day. Present condition, prolapsus uteri, obstinate constipation, great loss of flesh and muscular power, appetite poor, stomach weak and tender on pressure ; also constant, dull, and sometimes sharp gastric pain. Headache, nervous prostration, loss of mind (general melancholia, with suicidal tendency, if chloral is withdrawn for a few days). No pain in neighborhood of joints, slight dimness of vision accompanying headache. No family history of morphine or opium habit or alcoholism. Extreme sensitiveness of sexual organs."

Its effect on menstruation is slight. It sometimes causes too free flow, sometimes a slight, continuous flow. I can find no facts supporting the belief that it, like morphine or opium, produces sterility. As menstruation seems to be unaffected it is probable that it does not do so.

* Letter to the Author. † June, 1871.

As bearing upon the effect of chloral on menstruation, the following letter is of interest. No replies seem to have been received.

To the Editor of the Lancet:—

SIR—Have any of your correspondents noticed one peculiarity in the action of hydrate of chloral—namely, its power entirely to check the catamenial discharge? I have three patients who are in the habit of taking the drug, for attacks of asthma, and who have assured me that its exhibition is attended with this one drawback, if taken at the onset of the monthly period. I have had no opportunity of verifying their statements, but imagine there must be some truth in them, as they are all three of them in different stations of life, do not know one another, and cannot have compared notes. Moreover, their statements were volunteered, and not elicited by leading questions.

Your obedient servant,   G. H. R. DABBS, M.D.

Newport, I. W., Sept. 21st, 1872.

CIRCULATION AND COMPOSITION OF THE BLOOD.

Aberrations in circulation are due to derangement of the nerves governing the vessels; in other words, the vaso-motor system.

Flushing of the face, fullness of the head, redness of the ears, congestion of the eyes, alternating hyperæmia and anæmia of the various organs, is the result. From this cause we have the intense splitting headache one day, vertigo the next.

Palpitation of the heart, a weak, intermittent, irregular pulse, cough due to lung hyperæmia, accompanied by excess of secretion, often follow. Heart murmurs are sometimes heard, chiefly in very anæmic patients. Palpitation and irregular action of the heart is due to three causes, in the majority of cases—irregular nerve supply, impoverished blood, and atony of the cardiac muscles. This is not so in all cases, for we sometimes get these same symptoms from single doses of chloral, referable wholly then, I think, to disturbed nerve supply.

Upon the blood itself chloral acts as a disorganizer. That it does so we know. How it does so it is impossible to tell. This deterioration is evidenced by its action on the skin, to be noticed fully further on ; by the tendency to hemorrhage from mucous membranes, purpuric spots beneath the skin, spongy gums, falling of the hair, loss of the finger and toe-nails, malnutrition of the muscles, brain, and nervous system, anæmia, dropsy, and the but too palpable evidences of a general breaking up of the whole system.

Ludwig Kirn,* who has contributed some very valuable observations on the long continued use of chloral, relates the following :—

"We come now to a fourth group of cases, in which both the quality of the symptoms and their greater or less extension in the organism indicate a distinct change in the composition of the blood.

"In this connection, the following cases, observed by Crichton Browne, may first be referred to ; their interesting phenomena justify a detailed report.

"Case i.—A woman, aged sixty-nine years, suffering from periodical mania, had twenty grains of chloral twice daily ; on the fourth day a redness was developed on the skin of the chest and shoulders, which did not vanish on pressure ; on the sixth day the eruption had extended over the whole trunk and limbs, livid spots and deep-red

* *Practitioner*, 1873.

patches alternating. The lips and the mucous membrane of the mouth were excoriated, the gums spongy, the tongue blistered and ulcerated, the breath fœtid. The general state was one of great depression; pulse 120. On the eleventh day the ulceration of the mouth had extended further; the lips were covered with crusts. The petechial eruption was diminished on the chest and abdomen; the spots were yellowish, with patches of white skin between them; the spots on the arm lost their redness later. On the fifteenth day there was a sort of general desquamation; fissures of the skin over the sacrum and in the neighborhood of the joints. From that time convalescence proceeded and ordinary health was restored.

" CASE 2.—A woman, aged forty-six years, suffering from cardiac disease, hemiplegia, and dementia, took fifteen grains of chloral, three times a day, with calming effect. On the nineteenth day of the treatment numerous purple-red spots appeared in the neighborhood of the left elbow; on the next day many similar spots were seen on the shoulders and forearms, which coalesced with the others. On the twenty-first day livid spots came on the face; the left arm swelled and became hard. On its surface appeared a multitude of minute points, of a much deeper color, which did not diminish on pressure. Next day there were dark purple spots and discolorations; some small, round, and circumscribed; others broad and irregularly shaped, on the legs and abdomen, and in stripes on either side of the vertebral column. Simultaneously with the petechia there was great prostration, tendency to somnolence, weakness and excitability of the pulse, sore lips, thickly coated tongue. On the twenty-third day the spots and discolored patches had extended in every direction, and the previously bright-red spots had assumed a deep-purple color. Finally, signs of lung congestion appeared, with gradual failure of power, and death, after several fainting fits, on the twenty-sixth day. At the autopsy numerous ecchymoses, of every shape and size, were observed, more or less, on all parts of the skin; the right lung was congested and œdematous; the heart dilated and its valves thickened; over the right central hemisphere there was a large arachnoid cyst, containing fluid blood.

"With the foregoing may be joined a case, related by Monkton, in which, after four days' administration of sixty grains of chloral daily, a rash resembling slight variola, with hemorrhagic purpura, appeared, and death occurred on the sixth day, by syncope.

"Finally, may be mentioned two patients of Pelman, in whom, after treatment with chloral, there were larger and smaller petechia over the whole skin; in one which proved fatal numerous petechia were found on the mucous membrane of the larynx and under the endocardium, and a hæmatoma on the right side in the skull, reaching to the base, the fluid contents of which gave evidence of their recent origin.

"I shall now relate a case observed by myself, which is yet more striking, from the multiplicity of its phenomena, which are of a kind, perhaps, to give us some clearer understanding of their origin. The case was observed by me at a time when the evil effects of chloral hydrate were not yet known; the medicine was, therefore, continued, in spite of the most multiplied symptoms, because chloral was not for a long time recognized as their cause; the affection was, therefore, followed up further than we should dare to do now, that our increased knowledge would oblige us to stop at an earlier stage. [Here follows a detailed report of the case, which need not be given]. If we may sum up the weightier symptoms of this case, we find a young, strong, personally healthy person suffering from uncomplicated mania, in whom, on the ninth day of chloral treatment, a rash appeared in the form of groups of red spots, which soon became confluent. On the twentieth day the temperature and pulse rapidly rose to a febrile pitch; three days later the temperature had reached 106.7°; large and repeated doses of quinine were given without result, and baths had only a temporary effect. Œdematous swelling of the face, cheeks, eyelids and ears now set in. During the whole course of the disease the skin, so far from returning to its natural appearance, was the seat, now of impetiginous, now of moist, now of scaly eczema and ichthyoses, so that the process of desquamation, instead of being short, as in the acute exanthemata, occupied many weeks, during which great sheaths of epidermis were cast off from all parts of the body.

"The profound lesions of the skin nutrition were evidenced in the later stages by a remarkable shedding of the hair, and a gradual falling off of all the nails of the hands and feet. The affection of the skin was accompanied by a similar one of the mucous membranes, first of the intestines, which kept up watery diarrhœa in spite of medicine, and then by a similar affection of the conjunctiva and the bronchi. From the sixth week of the disease onward a series of large abscesses formed on both arms, over the shoulders and armpits, which secreted a considerable quantity of pus. While these phenomena were occurring there had been for eight weeks a continuous fever, occasionally remitting, and then again running up to a temperature beyond 104°.

"The symptoms which we have now collectively described must be defined as chronic blood poisoning. We cannot, however, place this in any of the known groups; we have not to do with a pyæmia or septicæmia, nor with a metallic or vegetable poisoning, since none of the causes have been at work which would lead to these affections, nor do we observe their characteristic phenomena; still less did the affection which the medicine produced in this patient resemble puerperal mania. In fact, there was no other external cause except the administration of the chloral; this medicine, which in even much larger single dose produces no such effect, was for ten weeks given in nightly doses of forty to sixty or even seventy-five grains; sometimes in two doses daily. The symptoms began after a certain saturation had been produced by accumulation, to spread further and further, and finally to assume the complete picture of a chronic blood poisoning.

" The origin of the disease leads us thus, by exclusion, to conclusions which have a high degree of probability, and we are also in a position to adduce positive facts. If a glance be cast at the symptoms observed by ourselves and others, after a more or less continued administration of chloral, we meet with the greater part of the phenomena observed in our last case, especially the very various affections of the skin and mucous membranes, the alterations in vascular action, and finally, the profound alterations of the blood, which in some cases remind us of the phenomena of scurvy."

The characteristic feature in the morbid picture which we have given consists less in symptoms which in themselves are altogether new, than in the assemblage of the most heterogenous phenomena, which previously had only been observed singly, in one person, and in a most aggravated degree of intensity.

It is a peculiar fact, that such severe symptoms as are

here related, comparatively common in the early history
of the drug, are not now seen, save in very rare instances,
even in persons using larger quantities of the drug for a
much longer time.   There must have been some impurity
in the drug as then used.   It might be supposed, also,
that such effects would be more apt to occur in these
insane patients, much below par mentally and physically,
was it not well established that its prolonged use in like
cases, at the present day, does not produce the same effects,
save in occasional instances.*   The ulceration of the skin
about the finger and toe-nails, spoken of by so many ob-
servers.   Thus, Professor Nathan R. Smith, of Baltimore,†
reports two cases where this occurred.   Twenty-six similar
cases have been observed by Dr. James G. Kiernan,†
former Assistant Physician at the New York City Asylum
for the Insane.

From eighteen superintendents of insane and five physicians of inebriate asylums, from whom I have heard, not
one mentions this condition of skin about the edges of the
nails.   The only case of the kind reported to me is by Dr.
J. W. F. Webb, of Liberty, Miss., and consisted in simple
elevation of the skin at these points, without ulceration.

I think we must conclude that there was some impurity
of the drug used, although Dr. Kiernan avers that the
drug was that furnished by Powers & Weightman.

Disordered condition of the blood manifests itself in
other skin affections.   Judging from its effects in single or
small doses in certain persons who manifest an idiosyncrasy, would lead us to entertain the possibility of skin
affections from the long continued use of the drug.   An
artificial idiosyncrasy seems to be developed.

* Kane; "Chloral Hydrate in the Treatment of the Insomnia of Insanity and
Delirium Tremens." *N. Y. Medical Record,* Jan. 8 and 15, 1881.
   † Quoted by Mattison. Op. cit.

Single doses, or the use of moderate doses for a short time, has been found to produce erythema, eczema, papular, pustular, or macular eruptions ; also urticaria and herpes zoster. In a few well-authenticated instances, eruptions, exactly resembling and in some cases mistaken for those of measles and scarlatina, have been produced. They passed through regular gradations, finishing with the typical desquamation of the disease simulated. Further use of the chloral again produced the same disturbances.* It has been suggested that these effects are due to disorders of the digestion, as excessive acidity, etc., produced by chloral. This is disproved by the fact that in the majority of instances where they occur, the digestive organs are not disturbed.

Schüle† believes that the majority of these skin affections are due to nervous disturbance.

The following cases of hemorrhage are of interest in this connection :—

Dr. A. E. M. Rae (Edinburgh *Medical Journal*, Nov., 1871) relates a case of hysteria, where he gave chloral for months. Bleeding from the lungs became so frequent and violent that twice the patient's life was despaired of, and he had to abandon the use of the drug entirely. He refers also to a case by Dr. Husband, where fatal hemorrhage

---

* Kane ; "Some Peculiar Effects of Chloral Hydrate on the Eyes and Skin." *N. Y. Medical Record*, 1881.

To those interested in the further study of the action of chloral on the skin, 1 give the following references :—

Browne, Crichton, *Lancet*, Vol. 1, p. 440. Curschmann, *Deutsches Archiv f. Klin. Med.*, 1871, p. 139. Brochin, M., *Bull. Gen. de Therap.*, Feb. 15, 1880. Burman, J. Wilkie, *Lancet*, Mar. 16, 1872. Farquharson, *Braithwaite's Epitome*, Mar., 1880. Ingalls, Wm., *N. Y. Medical Record*, 1871, p. 91. Liebreich, O., *Lancet*, June 16, 1877. Mayer, *La France Medicale*, 1876. Martinet, *Thesè de Paris*, 1879. Neal, Breward, *Lancet*, Aug. 23, 1873. Smith, Nathan R., *N. Y. Medical Record*, 1871, p. 299. Kiernan, J. G.; Quoted by Mattison. Op. cit.

† *Allgemeine Zeitschrift f. Psychiatrie*, Bd. 28, Heft. 1.

13

from a fibrous tumor of the womb occurred in a patient who was taking chloral. Spencer Wells (*Medical Times and Gazette*, Sept. 18, 1869) has, however, given it in cases of cancer of the womb, without producing bleeding. Turnbull (Philadelphia *Medical and Surgical Reporter*, Aug. 31st, 1872) claims that it increases the menstrual flow. Dr. C. R. Cullen, of Richmond, Va., writes me that he has seen flooding follow its use. R. C. Shettle, Physician to the Royal Berkshire Hospital, believes that its use is dangerous in labor cases, owing to the likelihood of flooding afterwards. This is totally disproved by a mass of testimony sent me by my correspondents.*

Bleeding from the nose has been noted by Inglis (Edinburgh *Medical Journal*, Sept., 1877), by Mauriac, in three cases (*Gazette des Hopitaux*, 1870, p. 405) and Dr. F. Delmont, of San Buena Ventura, Cal. (by letter). This was a hysterical lady, who was using large quantities of chloral. After one of her spasms about an ounce of blood flowed from her mouth in a fine stream, as though thrown by a syringe.

MUSCULAR SYSTEM.

Upon this system the effects are decided, and are chiefly produced through the agency of the nervous system.† Trembling, spasm of isolated muscles and muscular fibres, convulsions, paralysis, loss of coördinative power, etc., have already been spoken of.

Certain symptoms, as flabbiness, paleness, atrophy and weakness, result from imperfect nutrition. By confining

* Kane, "Chloral in Obstetric Practice." *American Journal of Obstetrics*, March, 1881.

† This is conclusively proved by the experiments of both Labbée and Rajewsky, quoted by H. C. Wood, "Materia Medica and Therapeutics," Philadelphia, 1877, p. 318, 319.

these patients to the house a great deal it still further lowers the tone of the general health.

RESPIRATION.

This vital function is very seriously affected in certain advanced cases. Dyspnœa is the most prominent symptom. This is more marked if alcoholic stimulants are taken. The trouble is undoubtedly of purely nervous origin. It is usually accompanied by a slight cough, and excessive secretion of mucus. Ludwig Kirn * says:—

"An important symptom which we have noticed in a series of cases of the long continued use of chloral is an interference with respiration, which may remain slight and scarcely troublesome to the patient, or may become positive dyspnœa.

" This symptom was experimentally produced by the Swede, Hammersten, who observed severe dyspnœa in a cat that had taken chloral, and was briefly noticed by Jastrowitz, one of whose patients, while taking chloral, suffered from severe dyspnœa, with occasional cessation of breathing ; and it was finally completely described and explained by Schüle, who observed a patient who, after a long use of chloral, used regularly to suffer after meals from a sense of oppression, which made going up stairs extremely difficult, and even interfered with speech, although there was no chest disease to account for this. The symptoms persistently recurred in spite of all treatment, until the chloral was left off, when the oppression entirely disappeared. A similar chloral dyspnœa, though not so long continued, occurred in many cases observed by us, either with or without a rash, and a feeling of heaviness and anxiety. That the chloral dyspnœa does not always stop at the lower

* *Practitioner*, 1873.

degrees, but may proceed to the most severe and dangerous developments, is shown by the following observation communicated to me by an eminent physician. This gentleman was summoned in consultation by a lady prostrated by long suffering, who had of late suffered from attacks of extreme dyspnœa, which had increased to asphyxia. At the same time the face was swollen, the facial muscles paralyzed, and there were also all the signs of cerebral effusion.

"Every remedy had failed, and the patient seemed on the brink of the grave. The physician, therefore, recommended the discontinuance of a daily dose of forty-five grains of chloral which had been given as an hypnotic, whereupon all these highly alarming symptoms vanished, in an almost magical way; the cerebral disturbance ceased, and the respiration quickly resumed its normal type. The dyspnœa may be anatomically explained by analogy with the effects of chloral upon the skin and mucous membrane, by hyperæmia of the lungs, which is produced through the channel of the vaso-motor nerves.

"We find here a further confirmation of the assumption that chloral operates upon the vaso-motor centre and the medulla oblongata, and that its paralyzing influence extends thence to the peripheral branches of the affected nerves. This might also lead to a practical contra-indication of chloral in all morbid conditions where there is a tendency to congestions or stases of blood in the lungs."

In non-habituès it has been found by Fothergill* to produce dyspnœa where there is cardiac disease:—

"A patient was taken into the West London Hospital, with emphysema and aortic stenosis. In spite of rest, digitalis and ammonia, he was liable to attacks of dyspnœa,

* "Antagonism of Therapeutic Agents," p. 54.

which had come on *since* his admission into the hospital. On searching for an explanation, it was found that the house-surgeon had benevolently prescribed chloral for the sleeplessness complained of. This was at once stopped, and the attacks of dyspnœa never returned, though the man gradually sank.''

Dyspnœa is reported by thirteen of my correspondents as a symptom of the continued use of chloral.

On the throat and larynx the effect is one of decided irritation, attended by congestion and sometimes ulceration. The uvula is congested and œdematous, the epiglottis red and swollen, and the vocal cords sometimes congested also. Cases of this kind are reported by Kirn* and Chapman†.

THE EYES.

Upon the eyes the continued use of chloral produces a very decided and characteristic effect. Some persons not addicted to the use of chloral manifest a peculiar idiosyncrasy with reference to it. A single dose sometimes, more often a few doses, will produce severe conjunctivitis, occasionally accompanied by œdema of the subjacent parts, as also of the face. Photophobia is often marked. Abundant testimony as to this effect of chloral on the eyes may be found in my article upon that subject in the N. Y. *Medical Record* for 1881.

Conjunctivitis from the continued use of chloral is reported by forty-one of my correspondents, œdema by thirteen, and photophobia by thirteen.

Sixteen report ''weakness of vision,'' nine report ''double sight,'' one cataract, one sudden blindness disappearing on discontinuance of the drug, one ptosis, four amblyopia, and six asthenopia. Two cases of temporary blindness

* Op. cit.   † *Lancet*, 1871.

from chloral are also reported.   This was in non-habituès.
Keyser,* of Philadelphia, reported the case of a gentleman
accustomed to sixty and eighty grain doses of chloral, who
suddenly became blind.   Opthalmoscopic examination
revealed great retinal anæmia.   The drug was discontinued,
and in a few days sight was restored.

Schület demonstrated that in chloral takers the retina was
congested.   The same condition was observed by Bouchut‡
in the retina of children completely anæsthetized.

Burke Haywood,§ of North Carolina, observed an elderly
man, who, after some weeks' use of chloral, began to
complain of dimmed vision, which persisted and increased
till the drug was withdrawn, when it gradually disappeared.

The case of ptosis is reported by Dr. H. C. Bigelow, of
Washington, D. C., as follows :—

" Female, aged thirty years, married, spare build, nervous temperament, and active mind.   Grains thirty to
sixty, daily, for eighteen months.   Emaciated, hysterical,
constipated, flatulent.   Temper irritable.   Ptosis of the
right eye, commencing iritis and photophobia.   Ulceration of os uteri.   Had an orgasm while being examined.
Vaginismus ; sexual appetite strong, but dislikes her husband and marital intercourse.   Suffers from insomnia, loss
of memory, mental unsteadiness.   She ascribes all these
symptoms to chloral."

Dr. N. C. Husted, of Tarrytown, N. Y., reports to me
the case of a lady who has used the drug in from ten to
twenty-grain doses, for about eighteen months.   She is
troubled with partial amaurosis and excessive lachrymation.

* Quoted by Mattison. Op. cit.
† *Allgemenie Zeitschrift f. Psychiatrie*, Bd. 28, Heft 1.
‡ Quoted by Labbèe, *Bull, Gen. de Therap.*, 1869. T. 2, p. 758.
§ Quoted by Mattison. Op. cit.

Dr. Edward Bradley, of this city, reports to me the case of a lady who took twenty grains of chloral, three times a day, for some time.   Eyesight gradually failed, and she was soon obliged to wear glasses.   Her eyes were thoroughly examined, and nothing abnormal was found.   On discontinuing the chloral the eye trouble speedily disappeared.   The effects of the drug on the general system, especially the digestive apparatus, was very marked.   The habit was broken without trouble.

For a full discussion of the subject here presented in outline, the reader is referred to my article in the *Record*, already spoken of.

CHAPTER XI.

SYMPTOMS OF ABSTINENCE FROM CHLORAL — DOSES AND DANGERS — TREATMENT.

The symptoms incident to the abrupt withdrawal of chloral from those who have used it for a long time are rarely severe; never so severe as those attending the same procedure in opium or morphine habituès. Many chloral takers voluntarily intermit the use of the drug for weeks at a time, themselves.

One of the most prominent symptoms that occurs occasionally is the supervention of severe delirium, very like that of delirium tremens. Such a case is reported by Dr. Geo. F. Elliot.* The patient, a man, aged thirty-five years, had, however, taken fifteen grains of opium, daily, for many years. He had for a few weeks substituted chloral, taking 200 grains of this drug daily. On withdrawing the chloral all the phenomena of delirium tremens appeared. It subsided on the use of large doses of tartar emetic and opium. Similar symptoms are reported by many of my correspondents.

Flashes of heat, nervous prostration, palpitation of the heart, dyspnœa, insomnia, sometimes persistent, intense headache with vertigo, and neuralgic pains in the occipital region, are likewise common.

Pains in the limbs are usually found, but are not so severe as those occurring after stopping the use of morphine. They are speedily relieved by large doses of gelseminum. This drug was first used for the pains occurring

* *Lancet*, May 24, 1873.

in the limbs of chloral eaters, during the continuance of the habit, by Mr. Herbert M. Morgan.* It acted well. Baths, electricity, the cold pack, and the measures recommended for the treatment of the morphia habituès, should be used. There should be no gradual reduction, unless the patient is very anæmic and much debilitated. Quinine in twenty-grain doses is an excellent sleep producer in these cases. The delirium tremens is best treated by digitalis and bromide of potassium. Stimulants should be freely used *for the first three days*, and strychnia and iron be given in large and frequently repeated doses. Cod-liver oil, extract of malt, and a generous diet should be prescribed. Pepsine should be given in fifteen to thirty grain doses after each meal. The bowels should be kept gently moving, by some mild laxative and cold water enemata. Hemorrhage from the bowels, stomach or urinary passages is readily controlled by the homœopathic tincture of witch-hazel, in ten-drop doses.

After the first ten days of treatment, strenuous efforts should be made to improve the patient's health and mind by means of exercise, free diet, good reading, and pleasant conversation.

The conjunctivitis is best treated by mild astringent applications, as tea, or the following :—

℞. Acid tannici, ... gr.vj
   Sodæ biboratis, ... gr.xv
   Vin. opii, ... ℥j
   Glycerinæ, ... ℥j
   Aq., ... ℥ij.   M.
SIG.—Eye wash.

The eye troubles and the skin affections usually pass away, without interference, a few days or weeks after the discontinuance of the chloral.

* *British Medical Journal*, Sept: 2, 1876.

Restraint, full control, and a thorough search of the patient are necessary with these as with opium or morphine takers.

The practice employed by some physicians, of "tapering off" chloral eaters on small doses of opium or morphine, I consider unnecessary, and extremely dangerous, for these patients, as is well established, are prone to go from one habit to another, and the use of these drugs is placing needless temptation in their way.

PROGNOSIS.

Cure may be assured if proper control of the patient is had. As much depends on proper after treatment of these cases as in that of the morphine habit.

Relapses less often occur in these than in opium habitués.

Finally, then, the prolonged use of chloral is not so likely to form a habit; is not so thoroughly enslaving when formed; is less prone to endanger life in small, more prone to destroy mind and body in large doses, and is easier broken, than the opium and morphine habits.

DANGERS.

There is a certain peculiar danger attending the use of chloral that is comparatively rare among opium habitués, viz : that of death from an overdose ; death, also, from a dose that has previously been taken with safety.* Medical literature is filled with records of such cases, and instances where death almost occurred, the patient only being saved through the exertions of the physician called.

In some instances an overdose was accidentally taken ; in others the person is found dead, it being probable that he took no more than his customary dose, which, however,

* Kane; "Deaths from Chloral Hydrate." N. Y. *Medical Record*, Dec., 1880.

acted with unusual strength upon a system surcharged with the drug.

Here, for instance, is, supposably, one of these cases. It is sent me by Dr. P. C. Remondino, of San Francisco, Cal. : —

"Isaac H., barrister, aged fifty-six; about four years ago began taking chloral for sleeplessness, due, he was told, to cerebral anæmia; was then a portly gentleman ; hair and beard dark ; beard slightly tinged with gray; feeling the need of stimulus, began to drink quite freely, and also to take morphia; then soon followed the habitual use of chloral. He now takes as much as one hundred and eighty grains per day, in three to four doses. Sometimes takes ten grains of morphia *with one of the doses;* is more than usually nervous ; now is greatly emaciated ; hair and beard a bleached white; skin itches and has a hard, dry feel ; has small brownish spots, about the size of this **O**, that cast off a small scab of skin ; suffers with pain in wrists and knees; also elbows and ankles, but not so severely as in the wrists and knees; insomnia and loss of appetite. Bowels regular; mental faculties are active ; of course nothing like those he formerly possessed, as his physical debility makes him childish, but still he can use his mind and will, to a certain extent ; acts perfectly gentlemanly, although he is as stated. His physical state is that of the debility and tottering of a man of ninety."

A short time afterward the doctor writes: "The chloral eater died some few days ago, in Los Angelos. Was found dead in the water closet of his hotel. The despatch says, 'supposed cause apoplexy.' I think it was more likely cardiac asthenia."

Dr. R. F. Lewis, of Lumberton, N. C., writes me : "A prominent physician of this place who was intemperate in the use of spirits, morphine, etc., began the use of chloral instead, and for weeks or months was more or less under its influence. He died suddenly after using it in increased quantities the day and night before. No autopsy.''

Dr. A. R. Kilpatrick, of Navasota, Texas, sends me the following curious note:—

''About four or five years ago there was a doctor living at Port Hudson, West Feliciana Parish, La., named (I think) Harris, who wrote several papers for the *Med. and Surg. Reporter*, of Philadelphia. He wrote one or two papers especially on the use of chloral and on the chloral habit, and very impressively warned people about its use and abuse, and in less than a year after the publication of those pieces I saw his death announced, and that he had been a habitual consumer of chloral, and that it killed him.''

Here is another case of death from an overdose, in an habituè, contributed by Dr. S. Henry Dessan, of this city:—

'' The only case where I have known death to be in any way connected with the administration of chloral, was in a case of dipsomania in a hysterical female. I prescribed a combination of fifteen grains of chloral with thirty grains of bromide of potassium, to be repeated every two hours, until sleep was procured; the effect was obtained after three or four doses. About eight doses were given in the mixture. I ceased attending the case, and about a year after learned through the press that the patient had died from an overdose of chloral. On inquiring from the druggist who had prepared the prescription, I learned that the patient continued to use the medicine steadily after my visits ceased, and that for twenty-four hours before death she had used two bottles of the mixture, or in other words

four ounces of chloral with one ounce of bromide of potassium.''

Two cases of chloral habituation in men past middle
age are reported by Dr. C. A. Bryce of Richmond, Va.,
where death occurred from *symptoms resembling apoplexy.*

THE HASHISCH HABIT.

CHAPTER XII.

HASHISCH INTOXICATION.

A common practice in some of the far Eastern countries
—hashisch taking—is comparatively rare among the people
of civilized nations. Here, as there, the practice is not one
of steady, daily intoxication with this drug, but it, more
like alcohol, is resorted to at certain times, when the system
seems especially to crave it, or the temptation is offered.
In this it differs materially from the practice of opium or
morphia taking. In point of continual craving, we might,
I think, arrange these drugs in the following order: Morphine or opium, chloral, hashisch, alcohol.

It would seem that, as the more intense is the daily or
hourly craving for a stimulant or narcotic, the easier it is
to permanently destroy that craving when the habit is once
broken. Thus a short struggle of from four to eight days
will, in the majority of instances, cure the opium patient,
while with alcohol or hashisch, less so with chloral, the
desire seems to be latent and to crop out at odd times, and
under peculiar circumstances. Once the desire is fully satisfied, then it remains quiescent for a shorter or longer
period, to again show itself in its original, or with increased
intensity, at a later date. Hence it is that it is so very
difficult to permanently cure dipsomania. With the opium

206

or morphine habitué, the desire at first, certain symptoms at a later date, come to the surface and demand a renewal of the drug saturation as soon as the effect of the last dose passes away.

We must differentiate between a diseased mental condition that imperatively calls for some narcotic or stimulant, and that craving for these substances that is only developed after their prolonged use, and which *did not exist before their use was begun.* Both conditions are those of disease; the one always existing, the other springing from the prolonged use of the substance to which they become addicted. Be it distinctly understood that where, throughout this book, I have used the word "habit," I have meant an *abnormal* appetite or condition, calling for the use of narcotics or stimulants, that either existed before or was produced by the use of the substance in question.

There are those who use hashisch steadily the year round, as many of our countrymen use alcohol; but this is due more to moral depravity than to any special morbid craving for the substance used.

Were we able to procure a thoroughly reliable extract of hemp in this country, and did physicians use it as freely, as carelessly, and in as large doses, as they are using opium, morphine and chloral, hashisch takers would be more common.

Known in English-speaking countries as *Indian Hemp* or *Cannabis Indica,* it is called *Hashisch* by the Arabians, *Gunjah* and *Churrus* by the inhabitants of India, *Bust* or *Sheera* by the Egyptians, *Dagga* or *Dacha* by the Hottentots, *El Mogen* by the Moors. This is the solid extract. *Bangue, Bang* or *Bendji* is the spirituous extract.*

Our pharmacopœia offers a tincture of Cannabis Indica,

* Calkins; "The Opium Habit," Phila., 1871.

every drachm of which represents fully three grains of the extract. Some of our manufacturing chemists prepare a fluid extract, and a fairly though not thoroughly reliable extract of hemp is manufactured by the English.

The English extract is that usually employed for medicinal purposes, and for the production of intoxication.

The only habituè that I have known was a woman, thirty-eight years of age, who consumed, daily, nine grains of the English extract. She would roll it up into a little lump, knead it for some time between the fingers, and then placing it in the bowl of a common clay pipe, partly filled with tobacco, light it, and inhale the smoke. This was done twice daily, about four and a half grains being used at a time. Sometimes she would go a week at a time—at least so she said—without using any; but I suspect that on those days when she did not smoke it, she used it by the stomach.

She was of an intensely nervous temperament, formerly addicted to the excessive use of alcoholic stimulants; of sallow complexion, dull eyes, pupils always *widely* dilated, pulse slow and irregular, occasionally intermitting a beat ; heart sounds feeble, body poorly nourished, skin dry, bowels usually loose, appetite poor, and urine scanty and high colored, but free from casts, albumen or sugar. She had been using the drug in this way about eighteen months, and found it necessary to occasionally increase her doses. She complained, especially in the morning, when waking from her almost cataleptic state, of intense pain in the *left* side of the head, and along the course of the sciatic nerve of the same side.

She began to use the drug through curiosity, having read of its peculiar effects, and being extremely desirous of finding something to supply the place of the alcohol, to which she had become a slave.

When not under the influence of the drug her intellect was dull and sluggish, and her temper, at times, bad and unreasoning. During the night, when most completely under the influence of hemp, her dreams were highly pleasurable; she seeming to live in a different world, a thought being answered by its accomplishment, a wish by its fulfillment; distances were traversed in a few seconds; feasts, marked by plenty and variety, the food on dishes of gold, studded with diamonds and other precious stones, were set before her. Everything was done on a scale of magnificence. At times the dreams partook of a highly lascivious character. She assured me that she seemed to be living a double life—the one the real, the other that produced by the hemp. In the latter the incidents of one night's dreams seemed to follow as regularly, and the characters to be as real, as the incidents and people of every day life.

There was one peculiarity: if she took a little more than her usual allowance of the drug, she found her dreams of an entirely different nature; not pleasant, but inexpressibly horrible, new faces and new scenes taking the place of the usual ones, the thread of her dream romance being suddenly snapped. The same thing occurred, though not so markedly, if she took less than the usual amount.

Before commencing the use of this drug she was in fair health, stout, and when not under the influence of liquor, bright and cheerful.

She passed entirely away from my observation, and I have never since been able to learn what became of her, though I heard once that she had died, how or when I do not know.

I once saw her in one of those deep sleeps produced by

14

hashisch, and noted that there seemed to be complete anæsthesia, deep snoring respiration, thirteen to the minute, dilated and *irregular* pupils, purplish congestion of the face and conjunctivæ, and a spasmodic twitching of the *left* eyelid that lasted all the time I was with her—two hours.

She was possessed of some money, and was very highly educated. She claimed to be the widow of an English army surgeon.

In the morning she rarely smoked all that was put in the pipe, and never enough to put her to sleep. Occasionally she added a little opium to the hemp. She was a mental and bodily wreck. Her gait was tottering, and sometimes she would be forced to go in a direction opposite to that in which she desired to move.

She expressed no desire to be broken of her vice, saying often that if she wished it she could stop without any trouble. I regret exceedingly that her temper and many peculiarities would not permit my studying her case more closely. The urine examined was obtained by catheterization during the semi-cataleptic spell already spoken of.

The mental effect of this drug has been variously described by different authors. Thus, Calkins* says :—

"The *mental condition* is an ideal existence, the most vivid, the most fascinating. Time and space both seem to have expanded by an enormous magnification ; pigmies have swelled to giants, mountains have grown out of molehills, days have enlarged to years and ages. De Moria in wending his way one evening to the opera house, seemed to himself to have been three years in traversing the corridor. De Saulcy having once fallen into a state of insensibility following upon incoherent dreamings, fancied he

* Op. cit. p. 323.

had lived meanwhile a hundred years. Rápidity as well as intensity of thought is a noticeable phenomenon. De Lucca, after swallowing a dose of the paste, saw, as in a flitting panorama, the various events of his entire life, all proceeding in orderly succession, though he was powerless in the attempt to arrest and detain a single one of them for a more deliberate contemplation. Memory is sometimes very singularly modified nevertheless, there being perhaps a forgetfulness, not of the object, but of its name proper, or the series of events that transpired during the paroxysm may have passed away into a total oblivion.

" The *normal mental condition* is that of an exuberant enjoyance rather than the opposite, that of melancholy and depression, though the transition from the one state to the other may be as extreme as it is swift. Oftener the subject is kept revolving in a delirious whirl of hallucinatory emotions, when images the most grotesque and illusions the drollest and most fantastic crowd along, one upon another, with a celerity almost transcending thought. (Mirza Abdúl Roussac.)

" *Command over the will* is maintainable, but temporarily only. As self-control declines the mind is swayed by the mere fortuitous vagaries of the fancy ; and now it is that the dominant characteristic or mental proclivity has its real apocalypsis. The outward expression may reveal itself under a show of complacency and contentment in view of things around, or suspicion, distrust, or querulousness of disposition may work to the surface, or may be, a lordly hauteur that exacts an unquestioning homage from the ' profanum vulgus' by virtue of an affected superiority over common mortals, is the ruling idea of the hour ; or peradventure, the erotic impulses may, for the time, over- · shadow and disguise all others.

"Amid the ever-shifting spectacular scene the *sense of personal identity* is never perhaps entirely lost, but there does arise in very rare instances the notion of a duality of existence; not the Persian idea precisely, that of two souls occupying one and the same body in a joint stock association, as it were (the doctrine as alluded to by Xenophon, in the story of the beautiful Panthea), but rather the idea of one and the same soul in a duplication or bipartation else, and present in two bôdies."

Many persons who have put themselves once or twice under the influence of this drug claim that no such pleasant effects, but rather torturing and horrible conditions are produced. The results when the drug is taken in this way, like those produced by tobacco on boys who smoke for the first time, should not be taken as a true estimate of the results obtained by the continued use of either.

A curious, interesting and valuable experiment was made upon himself by Dr. H. C. Wood, of Philadelphia, who is especially qualified to undertake and record the results of such an observation.

He says*:—

"When given in full doses, cannabis Indica produces a feeling of exhilaration, with a condition of reverie, and a train of mental and nervous phenomena, which varies very much according to the temperament or idiosyncrasies of the subject, and very probably also, to some extent, according to the nature of his surroundings. The sensations are generally spoken of as very pleasurable; often beautiful visions float before the eyes, and a sense of ecstacy fills the whole being; sometimes the venereal appetites are greatly excited; sometimes loud laughter, constant giggling, and other indications of mirth are present. Some years since,

* *Materia Medica and Therapeutics*, Phila., 1877, p. 226.

in experimenting with the American extract, I took a very large dose, and in the essay upon the subject (*Proceedings of the American Philosophical Society*, 186$, vol. XI. p. 226), the result was described as follows:—

" 'About half-past four P.M., September 23, I took most of the extract.· No immediate symptoms were produced. About seven P.M. a professional call was requested, and, forgetting all about the hemp, I went out and saw my patient. While writing the prescription I became perfectly oblivious to surrounding objects, but went on writing, without any check to or deviation from the ordinary series of mental acts connected with the process, at least that I am aware of. When the recipe was finished, I suddenly recollected where I was, and, looking up, saw my patient sitting quietly before me. The conviction was irresistible that I had sat thus many minutes, perhaps hours, and directly the idea fastened itself that the hemp had commenced to act, and had thrown me into a trance-like state of considerable duration, during which I had been stupidly sitting before my wondering patient. I hastily arose and apologized for remaining so long, but was assured I had only been a very few minutes. About seven and a half P.M. I returned home. I was by this time quite excited, and the feeling of hilarity now rapidly increased. It was not a sensuous feeling, in the ordinary meaning of the term; it was not merely an intellectual excitation; it was a sort of *bien-être*—the very opposite to *malaise*. It did not come from without; it was not connected with any passion or sense. It was simply a feeling of inner joyousness; the heart seemed buoyant beyond all trouble; the whole system felt as though all sense of fatigue were forever banished; the mind gladly ran riot, free constantly to leap from one idea to another, apparently unbound from its ordinary laws. I was disposed to laugh; to make comic gestures; one very frequently recurrent fancy was to imitate with the arms the motions of a fiddler, and with the lips the tune he was supposed to be playing. There was nothing like wild delirium, nor any hallucinations that I remember. At no time had I any visions, or at least any that I can now call to mind; but a person who was with me at that time states that once I raised my head and exclaimed, 'Oh, the mountains! the mountains!' While I was performing the various antics already alluded to, I knew very well I was acting exceedingly foolishly, but could not control myself. I think it was about eight o'clock when I began to have a feeling of numbness in my limbs, also a sense of general uneasiness and unrest, and a fear lest I had taken an overdose. I now constantly walked about the house; my skin, to myself, was warm—in fact, my whole surface felt flushed; my mouth and throat were very dry; my legs put on a strange, foreign feeling, as though they were not a part of my body. I counted my pulse and found it one hundred and twenty, quite full and strong. A foreboding, an undefined, horrible fear, as of impending death, now commenced to creep over me; in haste I sent for medical aid. The curious sensations in my limbs increased. My legs felt as though they were waxen pillars beneath me. I remember feeling them with my hand and finding them, as I thought, at least, very firm, the muscles all in a state of tonic contraction. About eight o'clock I began to have marked 'spells'—periods when all connection seemed to be severed between the external world and myself. I might be said to have been unconscious during these times, in so far that I was oblivious to all external objects, but on coming out of one it was not a blank, dreamless void, upon which I looked back, a mere empty space, but rather a period of active but aimless life. I do not think there was any connected thought in them; they seemed simply wild reveries, without any binding cord—each a mere chaos of disjointed ideas. The mind seemed freed from its ordinary laws of association, so that it passed from idea to idea, as it were, perfectly at random. The duration of these spells, to me, was very great, although they really lasted but a few seconds to a minute or two. Indeed, I now entirely lost my power of measuring time. Seconds seemed hours; minutes seemed days; hours seemed infinite. Still I was perfectly conscious during the intermissions between the paroxysms. I would look at my watch, and then, after an hour or two, as I thought, would look again, and find that scarcely five minutes had elapsed. I would gaze at its face in deep disgust, the minute-hand seemingly motionless, as though graven in the face itself; the laggard second-hand moving slowly, so slowly it appeared a hopeless task to watch during its whole infinite round of a minute, and always would I give it up in despair before the sixty seconds had elapsed. Occasionally, when my mind was most lucid, there was in it a sort of duplex action in regard to the duration of time. I would think to myself, it has been so long since a certain event—an hour, for example, since the doctor came; and then reason would say, no, it has been only a few minutes; your thoughts or feelings are caused by the hemp. Nevertheless, I was not able to shake off this sense of the almost indefinite prolongation of time, even for a minute. The paroxysms already alluded to were not accompanied with muscular relaxtion. About a quarter before nine o'clock I was standing at the door, anxiously watching for the doctor, and when the spells would come on I would remain standing, leaning slightly, perhaps, against the doorway. After awhile I saw a man approaching, whom I took to be the doctor. The sound of his steps told me he was walking very rapidly, and he was under a gas-lamp, not more than one-fourth of a square distant, yet he appeared a vast distance away, and a corresponding time approaching. This was the only occasion in which I noticed an exaggeration of distance; in the room it was not perceptible.   My extremities now began to grow cold, and I went into the house.   I do not remember further, until I was aroused by the doctor shaking or calling me.   Then intellection seemed pretty good.   I narrated what I had done and suffered, and told the doctor my opinion was that an emetic was indicated, both to remove any of the extract still remaining in my stomach, and also to arouse the nervous system.   I further suggested our going into the office, as more suitable than the parlor, where we then were.   There was at this time a very marked sense of numbness in my limbs, and what the doctor said was a hard pinch produced no pain.   When I attempted to walk upstairs my legs seemed as though their lower halves were made of lead.   After this there were no new symptoms, only an intensifying of those already mentioned.   The periods of unconsciousness became at once longer and more frequent, and during their absence intellection was more imperfect, although when thoroughly aroused I thought I reasoned and judged clearly.   The oppressive feeling of impending death became more intense.   It was horrible.   Each paroxysm would seem to have been the longest I had suffered; as I came out of it a voice seemed constantly saying, 'You are getting worse; your paroxysms are growing longer and deeper; they will overmaster you; you will die.'   A sense of personal antagonism between my will power, as affected by the drug, grew very strong. I felt as though my only chance was to struggle against these paroxysms—that I must constantly arouse myself by an effort of will; and that effort was made with infinite toil and pain. I felt as if some evil spirit had control of the whole of me, except the will power, and was in determined conflict with that, the last citadel of my being. I have never experienced anything like the fearful sense of almost hopeless anguish and utter weariness which was upon me. Once or twice during a paroxysm I had what might be called nightmare sensations; I felt myself mounting upward, expanding, dilating, dissolving into the wide confines of space, overwhelmed by a horrible, rending, unutterable despair. Then, with tremendous effort, I seemed to shake this off, and to start up with the shuddering thought, next time you will not be able to throw this off, and what then? Under the influence of an emetic I vomited freely, without nausea, and without much relief. About midnight, at the suggestion of the doctors, I went up-stairs to bed. My legs and feet seemed heavy, I could scarcely move them, and it was as much as I could do to walk with help. I have no recollection whatever of being undressed, but am told I went immediately to sleep. When I awoke, early in the morning, my mind was at first clear, but in a few minutes the paroxysms, similar to those of the evening, came on again, and recurred at more or less brief intervals until late in the afternoon. All of the day there was marked anæsthesia of the skin. At no time were there any aphrodisiac feelings produced. There was a marked increase of the urinary secretion. There were no after effects, such as nausea, headache, or constipation of the bowels.

" The sense of prolongation of time which I experienced was to me very remarkable, but.is not uncommon in these cases.   It is evidently due to the immense rapidity of the succession of ideas.   The mind, without doubt, measures time by the duration of its own processes, and when an infinitude of ideas arise before it in the time usually occupied by a few, time becomes infinitely prolonged to the mind.   It is a lifetime in the minute.   A very common mental phenomenon, not yet mentioned, is a condition of double consciousness, a sense of having two existences, of being at the same time one's self and somebody else.''

Pleasurable as may be the stage of excitement or intoxication, fascinating as may be the dreams that follow, the evil effect upon the body is rapid and decisive.

Wasting of the muscles, sallowness of the skin, hebetude of the mind, interference with coördination, failure of the appetite, convulsive seizures, loss of strength, and idiotic offspring, seem, from all accounts, to be the uniform result of the long continued use of this drug.

CHAPTER XIII.

CONCLUSIONS.

A careful study of the facts presented in the foregoing chapters teach us several lessons well worth considering, and suggest certain cautions and reforms that are greatly needed.

From the abundant evidence upon this point I think we must conclude that the abuse or habitual use of narcotics is steadily upon the increase, especially the subcutaneous use of morphine ; that these drugs are, in the majority of instances, first taken to relieve pain, and not for simple gratification of a morbid appetite ; and that the drug used and the manner of using it is in consonance with the prevalent medical practice of the time in which the habitué lives.

There are two classes especially blamable for this—the physicians and the druggists. In the early history of the use of the hypodermic syringe the danger of contracting the habit through its frequent use was not recognized, and the physician was not then to blame. At the present time, however, knowing fully the dangers incident to its use, the physician is criminally careless in placing the instrument in the hands of the patient or her friends for their use. If he does not appreciate the full extent of the danger, he is culpably ignorant, and certainly deserving of punishment.

The deaths, and dangerous accidents, and the spread of the continued use of narcotics, is due, to a great extent, to the druggists, who, in many cases, sell the drug without a physician's prescription, and without any reasonable excuse on the part of the patient, in direct violation of the law. Chloral is sold to men just recovering from a spree ; prescriptions containing large amounts of these drugs are renewed for patients for whom they were not originally given ; the druggist himself often prescribes a mixture of chloral, morphine and bromide of potassium, for repentant drunkards, or for patients suffering from insomnia.

When spoken to about this matter, they coolly excuse their practices with the remark that "if we don't do it, some other druggist will; and why should we lose the money." The laws relating to the sale of poisons are loose and inefficient, the practice rotten, and the statute really a dead letter. Dangerous and even fatal consequences* are not, then, so much to be wondered at.

Another matter in this connection needs attention, viz: the lying pretensions of a few charlatans, notably in the West, who, by specious advertisements and deceitful lies, induce the victims to these habits to buy their medicines, or come under their care for treatment. Their so-called specifics are simply preparations of opium or morphine, and their treatment is based upon the plan of *substituting one form of the drug for another.*

These sharpers are utterly without conscience, and do not scruple to prey upon and undermine the health of their victims, in order to gain a few dollars. It is about time that the people found out that honest, honorable and trustworthy physicians, who have only the good of the patient at heart, do not advertise. It is a shameful fact that the religious press tolerates the advertisements of these charlatans in their columns. As a rule, the vilest advertisements are to be found in these newspapers. Owing to the moral

* Kane ; " Deaths from Chloral." N. Y. *Medical Record*, Dec. 25, 1880, Jan. 1, 1881.

weight supposed to be carried by these sheets, owing to their large circulation among the people, who look upon every word therein contained as truth, these announcements and endorsements do the people an infinite amount of harm. Can it be that the financial "backers" of these papers overrule the scruples of the religious editor? If so, while a good investment financially, it must be a very poor one morally.

I have emphasized the fact that the continued use of chloral is not so liable to end in the formation of a habit, as is the prolonged use of morphia or opium; not that physicians may exercise less care and discrimination in its employment, for the danger is sufficiently great, but simply to refute the statements of some men who are gone wild upon the subject of habituation and inebriety, and who suggest measures for reform, and plans for restraint and treatment, as impracticable and impossible as their statements are whimsical and truthless.

Finally, be it distinctly understood, that many of the symptoms enumerated as occurring in both the morphine and chloral habitués, but especially the latter, are only found where the drug has been used in large amount, or for a long time. Every symptom will, moreover, be modified somewhat by the systemic peculiarities of each patient.

The "mixed" habits, so called, where patients are using two or more narcotics at one time, have not been discussed separately, as they possess no distinctive characters, and the physician who understands the prominent points of each will have no trouble in detecting and treating these cases.

INDEX.

SELECT LIST OF BOOKS

FROM THE CATALOGUE OF

MR. PRESLEY BLAKISTON,

1012 Walnut Street, Philadelphia.

FOR GENERAL AND SCIENTIFIC READERS.

☞ *Any of the following books will be sent, postpaid, upon receipt of the price, or they will be found in the stock of most booksellers throughout the United States and Canada.*

ON SLIGHT AILMENTS. Their Nature and Treatment. By Lionel S. Beale, M.D. Large 12mo. Cloth. Price $1.75.

Among civilized nations a perfectly healthy individual seems to be the exception rather than the rule ; almost every one has experienced very frequent departures, of one kind or another, from the healthy state ; in most instances these derangements are slight, though perhaps showing very grave symptoms, needing a plain but quick remedy.

CONDENSATION OF CONTENTS.

The Tongue in Health and Slight Ailments, Appetite, Nausea, Thirst, Hunger, Indigestion, its Nature and Treatment, Dyspepsia, Constipation, and its Treatment, Diarrhœa, Vertigo, Giddiness, Biliousness, Sick Headache, Neuralgia, Rheumatism, on the Feverish and Inflammatory State, the Changes in Fever and Inflammation, Common Forms of Slight Inflammation, Nervousness, Wakefulness, Restlessness, etc., etc.

OPINIONS.

" It abounds in information and advice, and is written for popular use."— *Philadelphia Bulletin.*

" Singularly clear and felicitous."—*Sanitarian.*

" A valuable work for the family library."—*Boston Transcript.*

" A thoroughly practical work for non-professional as well as medical readers."—*Cincinnati Gazette.*

" It belongs to that class of works to which high commendation belongs." —*New Haven Palladium.*

" A scholarly book, singularly free from all technicalities, and almost as valuable to general readers as to members of the profession."—*Chicago Inter-Ocean.*

" Clear, practical, and a valuable instructor and guide."—*Baltimore Gazette.*

" The advice given as to treatment is so excellent, that no student or young practitioner should neglect to read it."—*Med. and Surgical Reporter.*

" In a very important sense, a popular book."—*Chicago Advance.*

" An admirable treatise upon the minor ills which flesh is heir to."— *Springfield Republican.*

1

BRIGHT'S DISEASE. How Persons Threatened or Afflicted with this Disease Ought to Live. By J. F. Edwards, M.D. 16mo, 96 pages. Cloth. Price 75 cents.

The author gives, in a readable manner, those instructions in relation to Hygiene, Clothing, Eating, Bathing, etc., etc., which, when carried out, will prolong the life of those suffering from this disease, and a neglect of which costs annually many lives.

WHAT IS SAID OF IT.

" Every one should read this excellent little volume, in which Dr. Edwards describes and defines the disease."—*Providence Journal.*

" This little book is prepared, not in the interest of the doctor, but of the sufferer."—*Louisville Christian Observer.*

" A very valuable work."—*New York Commercial Advertiser.*

" Plainly written, and ought to be of great use."—*Philadelphia Public Ledger.*

" What should be done and avoided are clearly shown, and the information communicated is of general interest."—*Albany Journal.*

" Plain and straightforward."—*Baltimore Sun.*

" An admirable and much needed book."—*Catholic Mirror, Baltimore.*

" A remarkably able and useful treatise upon an obscure and vital subject."—*North American.*

"Should be read carefully by every one."—*The Voice, Albany, N. Y.*

"It encourages the sufferer as well as instructs him."—*Congregationalist.*

" An intelligent work."—*Toledo Blade.*

" A clear statement of some of the rules of life, which will insure the longest lease of life, and the greatest measure of health."—*Providence Press.*

" A satisfactory treatise."—*Indianapolis Sentinel.*

"Of especial interest and importance, and should be universally known." --*Lutheran Observer.*

" Will be eagerly welcomed by thousands of people. The malady is one of a peculiarly insidious character, and it may be asserted with confidence that this book will be a very valuable one for medical men ae well as laymen. The book is written in good, plain English, and with clearness."—*Stoddart's Review.*

" Simple, practical directions that can be easily obeyed."—*Bookseller and Stationer.*

" We are glad to call attention to this volume."—*Southern Churchman.*

" The considerations presented in this little volume are of the greatest moment."—*N. E. Journal of Education.*

"This volume is especially commended to those in whose behalf it is written."—*Chicago Evening Journal.*

"Contains a deal of useful information."—*Rural New Yorker.*

" Physicians, as well as laymen, will find the work interesting, and will obtain many valuable hints as to the proper hygiene to be observed in this disease."—*Medical News, Cincinnati.*

" To those for whom it is designed, this manual can hardly fail to be a God-send."—*Buffalo Courier.*

BY THE SAME AUTHOR.

CONSTIPATION. Plainly Treated and Relieved without the Use of Drugs. By Joseph F. Edwards. 16mo. Cloth. Price 75 cents.

THE MANAGEMENT OF CHILDREN in Health and

Disease. By Mrs. Amie M. Hale, M.D. A book for Mothers. Second Edition. 12mo. Cloth. Price 50 cents.

THE PRESS COMMEND IT AS FOLLOWS:

"Altogether, it is a book which ought to be put into every baby basket, even if some lace-trimmed finery is left out, and should certainly stand on every nursery bureau."—*Philadelphia Public Ledger.*

"Marked by good sense, simplicity and helpfulness in an unusual degree."—*Boston Journal.*

"Admirable common-sense advice, which mothers would do well to have."—*Southern Churchman.*

"Contains invaluable instruction."—*Evening News, Detroit.*

"The importance of this book cannot be over estimated."—*N. E. Journal of Education.*

"A work for mothers, full of wisdom."—*Congregationalist.*

"Ought to be the means of saving many a young life."—*Philadelphia Inquirer.*

"Abounds in valuable information."—*Therapeutic Gazette.*

"Emphatically a book for mothers, and cannot fail to be useful to all who read it."—*Indiana Farmer.*

"Admirably simple, clear, sensible, and safe in its teachings."—*Friends' Review.*

"It should be upon every household table."—*Nashville Journal of Med. and Surg.*

BIBLE HYGIENE; or, Health Hints. By a Physician. This

book has been written, first, to impart in a popular and condensed form the elements of Hygiene. Second, to show how varied and important are the Health Hints contained in the Bible, and third, to prove that the secondary trendings of modern philosophy run in a parallel direction with the primary light of the Bible. 12mo. Cloth. Price $1.25.

NOTICES OF THE PRESS.

"The anonymous English author of this volume has written a decidedly readable and wholesome book."—*Philadelphia Press.*

"The scientific treatment of the subject is quite abreast of the present day, and is so clear and free from unnecessary technicalities that readers of all classes may peruse it with satisfaction and advantage."—*Edinburgh Medical Journal.*

HEALTH AND HEALTHY HOMES. A Guide to Personal

and Domestic Hygiene. By George Wilson, M.A., M.D., Medical Officer of Health. Edited by Jos. G. Richardson, Professor of Hygiene at the University of Pennsylvania. 12mo. Cloth. 314 pp. Price $1.50.

CONTENTS.

CHAP.	PAGE.	CHAP.	PAGE
I. Introductory.	17	VI. Exercise, Recreation and Training,	187
II. The Human Body,	33		
III. Causes of Disease,	66	VII. Home and Its Surroundings, Drainage, Warming, etc.,	221
IV. Food and Diet,	119		
V. Cleanliness and Clothing,	169	VIII. Infectious Diseases and their Prevention,	269

"A most useful and, in every way, acceptable book."—*New York Herald.*

THE AMERICAN HEALTH PRIMERS. Edited by W. W.
Keen, M.D. Bound in Cloth. Price 50 cents each.

The Twelve Volumes, in Handsome Cloth Box, $6.00.

I. **Hearing and How to Keep It.** With illustrations. By Chas. H. Burnett, M.D., of Philadelphia, Aurist to the Presbyterian Hospital, etc.

II. **Long Life, and How to Reach It.** By J. G. Richardson, M.D., of Philadelphia, Professor of Hygiene in the University of Pennsylvania.

III. **The Summer and Its Diseases.** By James C. Wilson, M.D., of Philadelphia, Lecturer on Physical Diagnosis in Jefferson Medical College.

IV. **Eyesight, and How to Care for It.** With Illustrations. By George C. Harlan, M.D., of Philadelphia, Surgeon to the Wills (Eye) Hospital.

V. **The Throat and the Voice.** With illustrations. By J. Solis Cohen, M.D., of Philadelphia, Lecturer on Diseases of the Throat in Jefferson Medical College, etc.

VI. **The Winter and Its Dangers.** By Hamilton Osgood, M.D., of Boston, Editorial Staff Boston *Medical and Surgical Journal.*

VII. **The Mouth and the Teeth.** With illustrations. By J. W. White, M.D., D.D.S., of Philadelphia, Editor of the *Dental Cosmos.*

VIII. **Brain Work and Overwork.** By H. C. Wood, Jr., M.D., of Philadelphia, Clinical Professor of Nervous Diseases in the University of Pennsylvania, etc.

IX. **Our Homes.** With illustrations. By Henry Hartshorne, M.D., of Philadelphia, formerly Professor of Hygiene in the University of Pennsylvania.

X. **The Skin in Health and Disease.** By L. D. Bulkley, M.D., of New York, Physician to the Skin Department of the Demilt Dispensary and of the New York Hospital.

XI. **Sea Air and Sea Bathing.** By John H. Packard, M.D., of Philadelphia, Surgeon to the Episcopal Hospital.

XII. **School and Industrial Hygiene.** By D. F. Lincoln, M.D., of Boston, Mass., Chairman Department of Health, American Social Science Association.

This series of American Health Primers is prepared to diffuse as widely and cheaply as possible, among all classes, a knowledge of the elementary facts of Preventive Medicine, and the bearings and applications of the latest and best researches in every branch of Medical and Hygienic Science. They are not intended (save incidentally) to assist in curing disease, but to teach people how to take care of themselves, their children, pupils, employés, etc.

They are written from an American standpoint, with especial reference to our Climate, Sanitary Legislation and Modes of Life; and in these respects we differ materially from other nations.

The subjects selected are of vital and practical importance in every-day life and are treated in as popular a style as is consistent with their nature. Each volume, if the subject calls for it, is fully illustrated, so that the text may be clearly and readily understood by any one heretofore entirely ignorant of the structure and functions of the body. The object being to furnish the general or unscientific reader, in a compact form and at a low price, reliable guides for the prevention of disease and the preservation of both body and mind in a healthy state.

The authors have been selected with great care, and on account of special fitness, each for his subject, by reason of its previous careful study, either privately or as public teachers.

NOTICES OF THE PRESS.

"As each little volume of this series has reached our hands we have found each in turn practical and well-written."—*New York School Journal.*

"This is volume No. 5 of the 'American Health Primers,' each of which *The Inter-Ocean* has had the pleasure to commend. In their practical teachings, learning, and sound sense, these volumes are worthy of all the compliments they have received. They teach what every man and woman should know, and yet what nine-tenths of the intelligent class are ignorant of, or at best, have but a smattering knowledge of."—*Chicago Inter-Ocean.*

"The series of American Health Primers, edited by Dr. Keen, of Philadelphia, and published by Presley Blakiston, deserves hearty commendation. These handbooks of practical suggestion are prepared by men whose professional competence is beyond question, and, for the most part, by those who have made the subject treated the specific study of their lives. Such was the little manual on 'Hearing,' compiled by a well-known aurist, and we now have a companion treatise, in *Eyesight and How to Care for It*, by Dr. George C. Harlan, surgeon to the Wills Eye Hospital. The author has contrived to make his theme intelligible and even interesting to the young by a judicious avoidance of technical language, and the occasional introduction of historical allusion. His simple and felicitous method of handling a difficult subject is conspicuous in the discussion of the diverse optical defects, both congenital and acquired, and of those injuries and diseases by which the eyesight may be impaired or lost. We are of the opinion that this little work will prove of special utility to parents and all persons intrusted with the care of the eyes."—*New York Sun.*

"The series of American Health Primers (now entirely completed) is presenting a large body of sound advice on various subjects, in a form which is at once attractive and serviceable. The several writers seem to hit the happy mean between the too technical and the too popular. They advise in a general way, without talking in such a manner as to make their readers begin to feel their own pulses, or to tinker their bodies without medical advice."—*Sunday-school Times.*

"*Brain Work and Overwork.* By Dr. H. C. Wood, Clinical Professor of Nervous diseases in the University of Pennsylvania. This is another volume of the admirable "Health Primers," published by Presley Blakiston. To city people this will prove the most valuable work of the series. It gives, in a condensed and practical form, just that information which is of such vital importance to sedentary men. It treats the whole subject of brain work and overwork, of rest, and recreation, and exercise in a plain and practical way, and yet with the authority of thorough and scientific knowledge. No man who values his health and his working power should fail to supply himself with this valuable little book."—*State Gazette, Trenton, N. J.*

WHAT TO DO FIRST in Accidents and Poisoning. By Charles W. Dulles, M.D. Illustrated. 18mo. Cloth. Price 50 cents.

PREFACE.

Whoever has seen how invaluable, in the presence of an accident, is the man or woman with a cool head, a steady hand, and some knowledge of what is best to be done, will not fail to appreciate the desirability of possessing these qualifications. To have them in an emergency one must acquire them before it arises, and it is with the hope of aiding any who wish to prepare themselves for such demands upon their own resources that the following suggestions have been put together.

OPINIONS.

"Of special practical value, and we commend it to all."—*Lutheran Observer.*
"Ought to be in everybody's hands."—*Times, Philadelphia.*
"Its usefulness entitles it to a wide and permanent circulation."—*Boston Gazette.*
"Just the thing for an emergency."—*Portland Transcript.*
"Of great practical value to the public."—*Wisconsin State Journal.*
"A complete guide for sudden emergencies."—*Philadelphia Ledger.*
"So plain and sensible that it ought to be introduced into every female seminary."—*Evening Chronicle, Pittsburgh.*
"The suggestions are of priceless value."—*The Traveler, Boston.*
"The book is invaluable."—*Providence Press.*
"A valuable addition to the domestic library."—*Boston Transcript.*
"Merits an extensive sale."—*St. Louis Courier of Medicine.*
"This is an exceedingly useful and well arranged little book."—*North American.*
"The instructions of this little book are necessary and timely."—*Christian Secretary.*
"Contains a variety of practical suggestions that no household can afford to do without."—*Contributor, Boston.*

EYESIGHT, GOOD AND BAD. The Preservation of Vision. By Robert Brudenel Carter, M.D., F.R.C.S. With many explanatory illustrations. 12mo. Cloth. Price $1.50.

PREFACE.

A large portion of the time of every ophthalmic surgeon is occupied, day after day, in repeating to successive patients precepts and injunctions which ought to be universally known and understood. The following pages contain an endeavor to make these precepts and injunctions, and the reasons for them, plainly intelligible to those who are most concerned in their observance.

WHAT IS THOUGHT OF IT.

"A very valuable book, and should be in everybody's hands."—*North American.*
"A valuable book for all who are interested in the best use and preservation of the vision."—*N. E. Journal of Education.*
"A compact volume, full of information to all classes of people."—*Bookseller and Stationer.*
"A comprehensive treatise, well calculated to educate the public."—*Kansas City Review.*
"Gives excellent advice."—*Chicago Journal.*
"To teachers particularly the book is of interest and importance."—*Educational Weekly.*

ON HEADACHES. Their Causes and Cure. By Henry G. Wright, M.D. Ninth thousand. 16mo. Cloth. Price 50 cents.

ON DEAFNESS, GIDDINESS and Noises in the Head. By Ed. Woakes, M.D. Illustrated. 2d edition. 12mo. Cloth. Price $2.50.

HYGIENE AND SANITARY SCIENCE. A Complete Handbook. 4th revised edition. Containing chapters on Public Health, Food, Air, Ventilation and Warming, Water, Water Analysis, Dwellings, Hospitals. Removal, Purification, Utilization of Sewage and Effects on Public Health, Drainage, Epidemics. Duties of Medical Officers of Health, etc. By Geo. Wilson, M.A., M.D. 12mo. Cloth. Price $2.75.

" A sound book, by a very competent writer."—*London Lancet.*

WATER ANALYSIS For Sanitary Purposes, with Hints for the Interpretation of Results. By E. Frankland, PH.D., D.C.L. Illustrated. 12mo. Cloth. Price $1.00.

" The name of the author is a sufficient testimonial to its accuracy and its practical value."—*Boston Journal of Chemistry.*

BY THE SAME AUTHOR.

HOW TO TEACH CHEMISTRY. Being Six Lectures to Science Teachers. Illustrated. 12mo. Cloth. Price $1.25.

THE ART OF PERFUMERY. The Methods of Obtaining the Odors of Plants and Instruction for the Manufacture of Perfumery, Dentifrices, Soap, etc. etc. By G. W. Septimus Piesse. 4th edition enlarged. 366 illustrations. 8vo. Cloth. Price $5.50.

POTABLE WATER. How to Form a Judgment on the Suitableness of water for Drinking Purposes. By Charles Ekin. 12mo. Price 75 cents.

HEALTH RESORTS of Europe, Asia and Africa. The result of the Author's own observations during several years of health travel in many lands. By T. M. Madden, M.D. 8vo. Cloth. Price $2.50.

THE OCEAN AS A HEALTH RESORT. A Handbook of Practical Information as to Sea Voyages. For the Use of Invalids and Tourists. By Wm. S. Wilson, M.D. Illustrated by a chart showing the ocean routes of steamers, and the physical geography of the sea. 12mo. Cloth. Price $2.50.

DWELLING HOUSES and Their Sanitary Arrangements and Construction. By W. H. Corfield. Illustrated. 12mo. Cloth. Price $1.25.

DRAINAGE FOR HEALTH. Easy lessons in Sanitary Science. By Joseph Wilson, M.D. Illustrated. 8vo. Cloth. Price $1.00.

SANITARY EXAMINATION OF WATER, AIR AND Food. By Cornelius B. Fox, M. D. 94 engravings. 12mo. Cloth. Price $4.00.

NUTRITION IN HEALTH AND DISEASE. A Contribution to Hygiene and Medicine. 3d edition. By J. Henry Bennett, M.D. 8vo. Cloth. Price $2.50.

HYGIENE AND CLIMATE in the Treatment of Consumption. 3d edition. By J. Henry Bennett, M.D. 8vo. Cloth. Price $2.50.

PRACTICAL HYGIENE. A Complete Manual for Army and Civil Medical Officers, Boards of Health, Engineers and Sanitarians. 5th edition. With many illustrations. By Ed. A. Parkes, M.D. 8vo. Cloth. Price $6.00.

VOCAL HYGIENE AND PHYSIOLOGY. With special reference to the Cultivation and Preservation of the Voice. For Singers and Speakers. With engravings. By Gordon Holmes, M.D. 12mo. Cloth. Price $2.00.

WORKS ON CHEMISTRY.

CHEMISTRY, INORGANIC AND ORGANIC. With Experiments and a Comparison of Equivalent and Molecular Formulæ. 295 Engravings. By C. L. Bloxam. 4th London edition revised. 8vo. Cloth. Price $4.00.

NOTES FOR CHEMICAL STUDENTS. Compiled from Fowne's and Other Manuals. By Albert J. Bernays, PH.D. 6th edition 16mo. Cloth. Price $1.25.

MEDICAL AND PHARMACEUTICAL CHEMISTRY. Synthetical, Descriptive and Analytical. 2d edition, completely rearranged and revised. By John Muter, M.A., M.D. Royal 8vo. Cloth. Price $6.00.

HANDBOOK OF MODERN CHEMISTRY, Organic and Inorganic. By C. Meymott Tidy, M.D. 8vo. 600 pages. Cloth. Price $5.00.

A PRIMER OF CHEMISTRY. Including Analysis. By Arthur Vacher. 32mo. Cloth. Price 50 cents.

COMMERCIAL ORGANIC ANALYSIS. Being a Treatise on the Properties, Proximate Analytical Examination, and Modes of Assaying the various Organic Chemicals and Preparations employed in the Arts, Manufactures, Medicine, etc. 8vo. Cloth. Price $3.50.

MISCELLANEOUS.

ON HOSPITALS AND PAYING WARDS throughout the World. Facts in Support of a Rearrangement of the System of Medical Relief. By Henry O. Burdett. 8vo. Cloth. Price $2.25.

COTTAGE HOSPITALS; Their Origin, Progress and Management. 2d edition, enlarged and illus. By Henry O. Burdett. $4.50.

DEFECTS OF SIGHT AND HEARING; Their Nature, Causes and Prevention. By T. Wharton Jones, F.R.S. 2d edition. 12mo. Cloth. Price 50 cents.

IMPERFECT DIGESTION; Its Causes and Treatment. By Arthur Leared, M.D., F.R.C.P. 6th edition. 12mo. Cloth. Price $1.50.

COMPEND OF DOMESTIC MEDICINE, and Companion to the Medicine Chest. By Savory and Moore. Illustrated. 12mo. Cloth. Price 50 cents.

HOW TO WORK WITH THE MICROSCOPE. A Complete Manual of Microscopical Manipulation. Containing full descriptions of all new processes of investigation, with directions for examining objects under the highest powers, and for photographing microscopical objects. By Lionel S. Beale, M.D. 5th edition, enlarged and containing over 400 illustrations, many being colored. 8vo. Cloth. Price $7.50.

MICROSCOPIC MOUNTING. A Complete Manual, with notes on the collection and examination of objects. By Jno. H. Martin. 2d edition. With 150 illustrations. 8vo. Cloth. Price $2.75.

SECTION CUTTING. A Practical Guide to the Preparation and Mounting of Sections for the Microscope. By Sylvester Marsh. Illustrated. 16mo. Cloth. Price 75 cents.

THE MICROSCOPE IN PRACTICAL MEDICINE. With full directions for examining, preparing and injecting objects, the various secretions, etc. By Lionel S. Beale, M.D. 4th edition. 500 illustrations. 8vo. Cloth. Price $7.50.

SEA AIR AND SEA BATHING; Their Influence on Health. A Guide for Visitors at the Seaside. By Chas. Parsons, M.D. 18mo. Cloth. Price 60 cents.